Corruption, Infrastructure Management and Public–Private Partnership

Public–Private Partnerships (PPP or 3Ps) allow the public sector to seek alternative funding and expertise from the private sector during procurement processes. Such partnerships, if executed with due diligence, often benefit the public immensely. Unfortunately, Public–Private Partnerships can be vulnerable to corruption. This book looks at what measures we can put in place to check corruption during procurement and what good governance strategies the public sector can adopt to improve the performance of 3Ps.

The book applies mathematical models to analyze 3Ps. It uses game theory to study the interaction and dynamics between the stakeholders and suggests strategies to reduce corruption risks in various 3Ps stages. The authors explain through game theory-based simulation how governments can adopt a evaluating process at the start of each procurement to weed out undesirable private partners and why the government should take a more proactive approach.

Using a methodological framework rooted in mathematical models to illustrate how we can combat institutional corruption, this book is a helpful reference for anyone interested in public policymaking and public infrastructure management.

Mohammad Heydari is an Iranian writer and researcher. Currently, he is a candidate for the 'National Young Talent Program,' and he works as the youngest associate professor at Southwest University in Chongqing, China.

Kin Keung Lai is currently a Professor at International Business School, Shaanxi Normal University, Xi'an, China.

Zhou Xiaohu is currently a professor at the School of Economics and Management, Nanjing University of Science and Technology.

Routledge Advances in Risk Management
Edited by Kin Keung Lai and Shouyang Wang

For more information about this series, please visit: www.routledge.com/
Routledge-Advances-in-Risk-Management/book-series/RM001

Corruption, Infrastructure Management and Public–Private Partnership

Optimizing through Mathematical Models

Mohammad Heydari, Kin Keung Lai and Zhou Xiaohu

Routledge
Taylor & Francis Group

LONDON AND NEW YORK

First published 2022
by Routledge
2 Park Square, Milton Park, Abingdon, Oxon OX14 4RN

and by Routledge
605 Third Avenue, New York, NY 10158

Routledge is an imprint of the Taylor & Francis Group, an informa business

© 2022 Mohammad Heydari, Kin Keung Lai and Zhou Xiaohu

British Library Cataloguing-in-Publication Data
A catalogue record for this book is available from the British Library

Library of Congress Cataloging-in-Publication Data
Names: Heydari, Mohammad, author. | Lai, Kin Keung, author. | Xiaohu, Zhou, author.
Title: Corruption, infrastructure management and public-private partnership: optimizing through mathematical models/Mohammad Heydari, Kin Keung Lai and Zhou Xiaohu.
Description: Abingdon, Oxon; New York, NY: Routledge, 2022. | Series: Routledge advances in risk management |
Includes bibliographical references and index.
Identifiers: LCCN 2021020304
Subjects: LCSH: Public-private sector cooperation. | Infrastructure (Economics)–Management. | Corruption.
Classification: LCC HD3871 .H49 2022 | DDC 658/.046–dc23
LC record available at https://lccn.loc.gov/2021020304

ISBN: 978-1-032-01119-6 (hbk)
ISBN: 978-1-032-01123-3 (pbk)
ISBN: 978-1-003-17725-8 (ebk)

DOI: 10.4324/9781003177258

Typeset in Galliard
by Deanta Global Publishing Services, Chennai, India

Contents

Illustrations

Figures

Tables

Preface

Public–Private Partnerships (PPPs, 3Ps, or P3s) provide an opportunity for the public sector to seek alternative funding and expertise from the private party to procure public infrastructure. Governments can't provide all the citizenry's infrastructure in the face of budget constraints and other competing demands for state resources. Moreover, the private sector is a better resource manager, and the government should concentrate on policymaking. Where 3Ps are put to good use, the benefits are immense. Unfortunately, however, 3Ps, like public procurement, could be prone to corruption. This is the case; whatever gains that 3Ps provides to reduce the infrastructure deficit may be eroded, as corruption could result in inflation of the cost of construction or facilities' rehabilitation. Second, a 3Ps process marred by corruption could result in the use of inferior construction materials. A large chunk of funds would be diverted to bribing public officials by the project company. Third, a corrupt process could compromise the integrity of officials charged with the responsibility to inspect and approve construction works. Fourth, attracting genuine investors to participate in privately financed infrastructure development will become impossible if the country's 3Ps process is tainted by corruption. This book argues for stringent measures to be put in place to check corruption in 3Ps procurement in the world and the adoption of a good governance strategy to improve the performance of 3Ps in the country.

Governments worldwide are increasingly turning to 3Ps to deliver essential goods and services by leveraging private-sector resources and expertise. To be successful, 3Ps need to strike a balance between many public- and private-sector objectives. These include reliable and cost-effective service, stable financial returns, fair and predictable markets, and sustainable allocation of risks throughout the partnership arrangement. At any step in the project's life, corruption can undermine the balance of the 3Ps arrangement.

Many countries, including developed and emerging economies worldwide, find it difficult to cope with their infrastructure funding needs (Authers, J., 2015). This is due to budget constraints and competing needs for state resources vis-à-vis the enormous costs involved in infrastructure procurement. As a result, the public sector seeks to partner with the private sector to tap capital and expertise. This collaboration between the public and the private parties is commonly referred to as a Public–Private Partnership3Ps.

There is no universally approved definition for the term 3Ps. The 3Ps model allows a government to transfer exclusive rights to a private-sector consortium to develop and operate an infrastructure facility under certain conditions for a fixed period, which is usually long-term (Liu, Q. et al., 2019). For this book, a 3Ps may be defined as:

> A long-term contract among a private section and a government entity for providing a public asset or service, in which the private section bears significant risk and management accountability and remuneration is linked to performance.
>
> (Bertelli, A.M., et al., 2019)

However, in this book's context, not every partnership between the public and private sectors is a 3Ps. For instance, even though the contractor is a private-sector contractor in a traditional public procurement, the relationship is not a 3Ps. In the case of a 3Ps, all phases of a project are awarded to a single consortium. Second, the funding structure of a 3Ps is different from that of traditional public procurement. A 3Ps is also different from a management or service contract. In the latter, a private company is engaged to manage a facility's operation or provide a service, for instance, cleaning or refuse collection. In a 3Ps, the private party is contracted to design, finance, build, and operate a facility for a period.

Thus, it is essential not to confuse a 3Ps with other forms of public procurement. 3Ps are long-term procurement contracts

> where the supplier takes responsibility for financing and building the infrastructure and managing and maintaining it. The DBFO model ('Design,' 'Build' 'Finance' and 'Operate'), the BOT model ('Build,' 'Operate' and 'Transfer') or the BOO ('Build,' 'Own' and 'Operate') are all standard contractual modes that feature bundling of building and operation in a single agreement with an only contractor (or consortium of contractors). 3Pss are used across Europe, Canada, the US, and several developing countries for the provision of public infrastructures and services in sections such as transport, energy, water, IT, prisons, waste management, schools, hospitals, and others.
>
> (Fleta-Asín, J., Muñoz, F., and Rosell-Martínez, J., 2019)

3Ps are not privatized. Under privatization, the government divests some or all its interests in a government-owned enterprise. A 3Ps government still maintains ownership of the facility even though it is transferred to the private-sector project company during the 3Ps contract tenure. A 3Ps is also not a traditional public procurement. In a 3Ps, the private sector is responsible for the facility's design, finance, operation, and maintenance. On the other hand, in public procurement, the relationship is not long-term, and the government is responsible for the design and funding of the project. Simultaneously, the private contractor is liable for the facility's construction (Bertelli, A. M., Mele, V., and Whitford, A. B., 2019).

One impediment to a successful 3Ps regime in any given territory is corruption (Ellis, J., 2019). According to Transparency International, 'corruption is the abuse of entrusted power for private gain' (Lambsdorff, J. G., 1999). It poses a formidable challenge for development in emerging economies such as China. If the government must adopt 3Ps, then the process must be free of corrupt practices. Good governance demands elements of transparency, equal treatment, and open competition. The lack of these elements is a source of worry for local and foreign investors.

Chapter 1 provides mathematical modeling on 3Ps based on game theory. Game theory can be described as 'the study of the mathematical algorithm of conflict and cooperation among rational and intelligent DMs' (Myerson, 1991). It can also be called 'conflict analysis' or 'interactive decision theory,' which can more accurately express the theory's essence. Still, game theory is the most famous and accepted name. In 3Ps, conflicts and strategic interactions among developers and governments are pervasive and play a crucial role in 3Ps projects' performance. Many complicated problems, such as opportunism, negotiations, competitive biddings, and partnerships, have challenged the wisdom of 3Ps participants. Therefore, game theory is very appealing as an analytical approach to study the interaction and dynamics between the 3Ps participants and suggest proper strategies for both governments and promoters.

Chapter 2 discusses governments' playing evaluating games by becoming the first mover and setting criteria for evaluating out unwanted developers. 'Evaluating' was utilized to refer to the market process studied in Rothschild and Stiglitz (1976), in which the data issue in the insurance model was similar to that in Spence's (1974) job market model as discussed previously in the introduction of game theory. In the evaluating game, the Receiver initiates the move, and the Sender responds. An effective evaluating scheme design can induce the Sender to reveal private data or types. For instance, in the job market analysis, the firm may make the first move by demonstrating a menu of contracts regarding the education levels and wage offers. The workers will then respond by choosing their preferred agreements. In the job market, a useful evaluating menu of contracts would induce worker H to choose the agreement with a high education level and a high wage, and worker L to choose a low education level and a low wage contract.

On the other hand, an ineffective evaluating scheme cannot induce the workers to reveal their types. For example, if the stipend level difference is too large, it may be optimal for worker L to imitate worker H to invest more in education and obtain a high stipend. If the stipend level difference is too small, it may be optimal for worker H to choose low education, thus eliminating the need to incur high education costs.

The usage of evaluating criteria can be regarded as governments' strategies for the implementation of 3Ps. The central problems and the game equilibrium concepts are very similar to those in signaling games. An adequate evaluating criterion must be that only those of the desired type can satisfy the requirement with reasonable costs. We argue that since 3Ps projects are generally initiated by governments, which often have particular goals to achieve, playing evaluating games seems more sensible and practical for governments than waiting for the developers or bidders to guess what the government wants to send out weak signals.

Chapter 3 will use the game theory modeling algorithm to analyze and build theories about some of the above challenging 3Ps issues. Through game theory modeling, a specific problem of concern is abstracted to a level that can be analyzed without losing the problem's critical components. Additionally, new insights or theories for the concerned problem are improved when the game models are solved. Whereas game theory modeling algorithms have been broadly applied to study problems in economics and other disciplines, only recently have these algorithms been used to research issues in engineering administration problems, including 3Ps (see Ho, 2001; 2005; 2006; Ho and Tsui, 2009; Ho and Tsui, 2010). In the 3Ps study, the author believes that there will be high potential to gain critical new insights and build new theories by applying this algorithm. These new theories will assist practitioners, including governments, developers, bankers, etc., to better cooperate with higher efficiency and effectiveness.

Public–Private Partnership projects are central to the UK government's public services procurement. 3Ps are competitively tendered, but a significant minority of projects attract insufficient bidders, making it challenging to demonstrate transparency and value for money. Based on an empirical study, this chapter of the current chapter posits an initial model of how private-sector bidders decide whether to bid for contracts; a distinction is drawn between two separate assessments made by potential bidders, the risk of bidding and the risk of the project. The chapter concludes that practitioners and scholars insufficiently understand the former evaluation.

Chapter 4 presents two evaluative criteria related to 3Ps infrastructure projects, i.e., private evaluative criteria and public evaluative criteria. These evaluative criteria are inversely associated; the higher the public profits, the lower the private surplus. To balance evaluative criteria in the multi-Creation decision this chapter improves a quantitative matching decision algorithm to select an optimal matching scheme for 3Ps infrastructure projects according to the hesitant fuzzy set (HFS) under unknown evaluative criterion weights. In the algorithm, Hesitant fuzzy set illustrates the evaluative criteria values, and groups fully consider multi-criterion information. The optimal model is built and solved via maximizing each criterion's total deviation so that the evaluative criterion weights are determined objectively. Then, the multi-criterion decision match-degree is counted, and a multi-objective optimization model is illustrated to select an optimal matching scheme with a min-max strategy. The results represent new insights and implications of the influence on evaluation criteria in the Multi-Creation decision.

In the second section of this chapter, we represent a 3Ps selection process. This process can be seen as a branch of the broader public investment management process – that is, at some point, a project is chosen as a potential 3Ps, and after that follows a 3Ps-specific process. However, this branching can happen at different points in the public investment process. For example, this could be:

Well-defined 3Ps processes generally mirror public investment management processes – for example, requiring approvals by the same bodies, as described further in *Institutional Accountabilities: Review and Approval* (Irwin, T., and Mokdad, T., 2010).

Many governments introduce criteria or checklists for 3Ps potential against which projects can be compared to support this evaluating process. 3Ps Potential evaluating elements in South Africa provide an example of such a checklist from the South Africa 3Ps Manual (Baporikar, N., 2020). Similar criteria may also be used for more detailed appraisal, as described in *Assessing Value for Money of the PPP*. At the evaluating stage, the idea is to check if the criteria are likely to be met to proceed to the next development level.

This chapter's last section presents a new approach to evaluating and selecting 3Ps projects. Due to the popularity of 3Ps projects and the observation that the evaluation and selection of projects can be considered as a multiple-criteria decision-making problem, this chapter proposes a new approach, namely, minimizing the total deviation from the ideal point, to decide on the weights associated with each criterion. We provide a numerical illustration to demonstrate the effectiveness of this approach.

In Chapter 5, 3Ps projects' options properties are identified by analyzing 3Ps projects' investment characteristics from a private-sector perspective. 3Ps project investment can be viewed as an EU call option, and thus a pricing approach for option value of 3Ps projects is set up according to the Black–Scholes option-pricing model. Through empirical analysis of the Beijing Metro Line 4 project, we conclude that a 3Ps project's total value considering the option value is higher than that obtained by the traditional valuation tools (net present value). Therefore, the application of real options theory in 3Ps project investment decisions, which fully considers the flexibility of the private sector, allows to make decision in the real solution and improves decision-making accuracy.

3Ps is a relationship between the public sector and the private sector established in a certain way to complete the investment and construction of public infrastructure together. The public sector (usually the government) would ensure private sector franchise law to expedite the construction process and efficient management afterward (Chung, D., Hensher, D. A., and Rose, J. M., 2010). Since the 3Ps model can combine both the public sector and private sector and improve the service quality while reducing construction investment, it has been widely used in foreign countries in the construction area since its first application in England in the 1990s.

3Ps projects are usually public infrastructure, which involves huge investment, unusually long construction periods, irreversibility, uncertainty, and competition. Thus, they are especially suitable for investigation with the real options algorithm.

In Chapter 6, we will explore the function of 3Ps in future urban infrastructure. 3Ps in urban development can be best described as a true partnership of public officials and private developers who 'have improvement ambitions that they could not complete alone' (Sagalyn, 2007, p. 8). In this form of public-(municipalities) and private-sector (private companies such as construction and property development firms, private banks, investment companies, etc.) coopera-tion, the aim usually is to accomplish a public task or a project by funding and operating based on a partnership in which the financial risks of the public sector are to be decreased. Limitations drive 3Ps in public funds to cover investment needs and increase public services' quality and efficiency (EC, 2003).

Finally, in the References chapter, we provide a database of selected papers to summarize the research done in the 3Ps area.

The study recognized four categories of factors that different industries can focus on to decrease response time in the face of catastrophic incidents of low probability and high impact: organizational structure, preparation, partnership, and reserve. The research derives new insights, presented as three propositions that relate the evaluation in 3Ps disruption to negative or potentially positive impact and reduce corrupt activities in the 3Ps.

We have intentionally limited our focus to articles that have already been pub-lished at the time of this writing as an attempt to understand the state of research at a fixed point in time. Naturally, in an emerging field like 3Ps, there is much research in the pipeline, and a new '*chapter*' is required periodically. We hope this is useful for today's scholars.

Objectives

From our perspective, the authors aim to connect the logic and probabilistic cal-culus used in 3Ps with questions of ethical challenges in the process of 3Ps and organizations.

Applying the framework of institutional corruption, the authors draw out some of these implications, focusing on the potential systemic effects of 3Pss – forces often neglected in the academic literature and, even more so, in policy discus-sions (Di Marco, M. et al., 2020). These systemic effects tend to be insidious and include erosion of public institutions' mission and integrity and trust and confidence in those institutions. Building on these concerns, we draw attention to the limitations of prevailing analytical approaches to 3Ps' ethics and suggest alternative ways of addressing the systemic ethical issues they raise.

Before embarking on this analysis, the authors provide a brief definition and taxonomy of Public–Private Partnerships and an overview of institutional corrup-tion as an analytical framework or lens for the ethical review of Public–Private Partnerships.

The main contribution of this book is summarized as follows. First, we for-mulate mathematical modeling on 3Ps based on game theory, in which each expert has specific but uncertain evaluation results on a set of candidate pro-jects. Therefore, the project evaluation problem with interval values is deemed a stochastic optimization problem. Second, we discuss how governments can play evaluating games by becoming the first mover and setting criteria for evaluating out unwanted promoters. The usage of evaluating criteria can be regarded as governments' strategies for the implementation of 3Ps. Third, the authors believe that, in 3Ps research, there will be high potential for gaining critical new insights and building new theories by applying this algorithm. These new theories will assist practitioners, including governments, developers, bankers, etc., to better cooperate with higher efficiency and effectiveness. Fourth, we discuss evaluation criteria; these evaluative criteria are inversely related; that is, the higher the public profits, the lower the private surplus. balance evaluation criteria, this chapter improves a quantitative matching decision model to choose an optimal matching scheme for 3Ps infrastructure projects according to the HFS under unknown evaluative criterion weights. Even though the classical project evaluation problem has been discovered mainly in the literature, this study's investigation is entirely new and of both academic and practical significance and value. Fifth, the options properties of 3Ps projects investment are identified by evaluating the investment characteristics of 3Ps projects from the perspective of a private party. Sixth, we explored the function of 3Ps in future urban infrastructure.

Tentative topics

Tentative topics to be covered in this course include:

- Public–Private Partnerships in different countries
- Corrupt practices in Public–Private Partnerships
- Challenge of corruption in Public–Private Partnerships
- Identifying, selecting, and evaluating the 3Ps based on mathematical models
- Simulation-based assessment of risks related to 3Ps
- Quantitative-qualitative mixed model risk allocation model for 3Ps projects
- Private and public interests in Public–Private Partnership agreements through optimization of equity capital structure
- Theory building through game theory modeling in 3Ps

Target audience

Senior-year undergraduate and master-level students who study risk management, project management, operational management, organizational behavior, quantitative management, or similar areas, and practitioners who wish to learn how to price, structure, trade, the products with more advanced models and techniques.

Note

The present book's results are significantly connected with the PhD dissertation of Mohammad Heydari, which was written at the Nanjing University of Science and Technology, entitled: 'A Cognitive Basis Perceived Corruption and Attitudes Towards Entrepreneurial Intention.' Supervisor: Professor Zhou Xiaohu, School of Economics and Management, Nanjing University of Science and Technology,

Nanjing, Jiangsu, China. For more information about this dissertation, you can contact Mohammad_Heydari@njust.edu.cn and njustzxh@njust.edu.cn. There are some questions contained in this book which clarify the purpose of further research. Also, it is necessary to mention that this book results from the ten years of research in different countries on 'human and organizational behavior.'

1 Risk management, corruption, and game theory in Public–Private Partnership

1.1 Introduction to the concepts of game theory and its modeling

Game theory can be specified as 'the study of mathematical models of conflict and cooperation among intelligent, rational decision-makers' (Myerson, 1991). It can also be defined as 'conflict analysis' or 'interactive decision theory,' which can express the theory's essence more accurately. Still, game theory is the most famous and approved name. In Public–Private Partnerships (3Ps), conflicts and strategic interactions among bidders and governments are widespread and play a crucial role in 3Ps projects' performance. Many complicated problems, such as opportunism, negotiations, competitive biddings, and partnerships, have challenged 3Ps participants' wisdom. Therefore, game theory is very useful as an analytical framework to research the interaction and dynamics among 3Ps participants and suggest proper strategies for both governments and bidders.

Throughout this chapter, the game theory modeling method will analyze and build theories on 3Pthe above challenging issues of 3Ps. Through game theory modeling, a specific problem of concern is abstracted to a level that can be studied without losing the problem's critical components. Additionally, new insights or theories for the concerned issue are developed when the game models are solved. Whereas the game theory modeling method has been broadly applied to research issues in economics and other disciplines, only recently has this algorithm been used to study problems in engineering management problems containing 3Ps (see Ho, 2001, 2005, 2006; Ho and Tsui, 2009; Ho and Tsui, 2010). In 3Ps research, the writers believe that there will be great potential to gain critical new insights and build new theories by applying them. These new theories will assist practitioners, including governments, developers, bankers, etc., in better cooperating with higher efficiency and effectiveness.

This chapter has two primary objectives. Our first aim is to illustrate the concept of game theory and the application of game theory modeling in 3Ps. Second, new insights and approaches concerning 3Ps from game theory modeling will be presented and discussed, particularly concentrating on the opportunism issues and the contingency view of 3Ps as a governance structure.

DOI: 10.4324/9781003177258-1

The following introduction of game theory fundamentals follows Gibbons (1992) and Binmore (1992). Here games will be categorized by whether players move sequentially and whether or not the information is complete. There are two primary games in terms of players' moving sequences. Complicated by the data owned by players, the games can be further categorized into static games of complete data, dynamic games of complete data, static games with incomplete data, and dynamic games of incomplete data. Fudenberg and Tirole (1991), Mas-Collel et al. (1995), and Myerson (1991) have excellent in-depth discussions on game theory. Readers who already know game theory might skip this section; otherwise, it is suggested to read this section to understand better the game theory modeling applications presented in the chapter.

Examples of a two-player game shall illustrate some essential concepts and definitions in game theory. General cases of n-players definitions are omitted for convenience. The first instance is the prisoner's dilemma, as shown in Figure 1.1. Two suspects are arrested and interrogated in separate cells. If both of them confess, they will be sentenced to jail for six years. If neither of them confesses, they will be sentenced to only one year. However, if one confesses and the other does not, the honest one will be rewarded via being released (in jail for zero years), and the other will be punished with nine years in jail. In Figure 1.1, the first number in each cell represents player 1's payoff, and the second number is for player 2. We use the left side of the table to describe player 1 and use the top of the table to represent player 2.

Figure 1.1 is a 'normal form representation' of a game that specifies the players in the game, the approaches available to the players, and each player's payoff for his approach. The normal form representation is generally used in representing a 'static game' in which they act simultaneously. More specifically, each player

| | Player 2 | |
	Confess	Not confess
Confess	(-6, -6)	(0, -9)
Not confess	(-9, 0)	(-1, -1)

(Player 1 labels the left side of the table.)

Figure 1.1 Prisoner's phenomenon.

does not know the other player's decision before making his own decision. If the payoff matrix, as can be seen in Figure 1.1, is known to all players, then the payoff matrix is 'common knowledge' to all players in a game. Also, the game players are supposed to be rational, i.e., it is supposed that they will always maximize their payoffs. This is one of the most critical assumptions in any economic analysis. If the players' reasonableness and the game structure, containing payoffs, are common knowledge, the game is called a game of 'complete data.' Conversely, if each player's possible payoff is privately known by himself only, it is a game with incomplete data or asymmetric data.

To answer how each prisoner will play/behave in this game, *NE*, one of the essential condition concepts in game theory, will be introduced. If a game theory makes a unique prediction about each player's choice, then it has to be that each player is willing to play the approach as predicted. Logically, this prediction should be the player's best response to the other player's expected strategy. No single player will want to deviate from the predicted plan; that is, the strategy is strategically stable or self-enforcing (Gibbons, 1992). This prediction following the above condition concept is called a 'Nash equilibrium' (NE). In the prisoner's dilemma, although the (Not confess, Not confess) may seem better for both players, it is unstable since every player wants to deviate from this condition to get extra benefit or avoid the other's betrayal. Any suspect who deviates from (Confess, Confess) will be hurt, and any suspect who deviates from (Not confess, Not confess) will be rewarded. Therefore, the only predicted approach that no player wants to deviate from is (Confess, Confess), and this is the NE in the prisoner's dilemma.

There will be multiple NEs, that is, the uniqueness of NE is not guaranteed. Fortunately, the existence of numerous equilibriums (Eqs.) will not be an issue. Much of game theory is an effort to recognize a compelling Eq in different classes of games to make the prediction appealing (Gibbons, 1992). For instance, the concepts of mixed strategy NE or focal point can resolve multiple NEs. A detailed treatment of this issue will not be introduced here. The Cournot model for the duopoly market in economics is an example of a static game of complete data.

Most of the analysis in this chapter will be using a dynamic game with complete or incomplete data. However, the previous introduction of the static game is essential because those concepts will be used repeatedly in other games and this chapter. In contrast to static games, players in a dynamic game move sequentially instead of simultaneously. Since the moves are sequential, it will be easier and more intuitive to demonstrate a dynamic game by a tree-like structure, called an 'extensive form' representation. In a dynamic game, suppose that the player who moves in a later stage can fully observe the previous player's moves and know his location in the game tree. This assumption is called the 'perfect information' assumption. In dealing with incomplete data games, the games will be transformed into games of 'imperfect information,' where the player who moves in a later stage cannot fully observe the previous player's moves.

We will use the following simplified market entry instance to demonstrate the concepts of game analysis. A new firm, New Inc., wants to enter a market

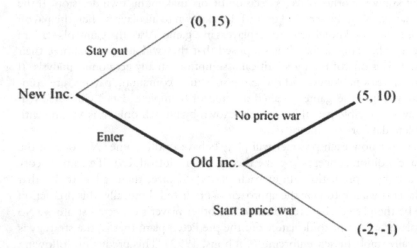

Figure 1.2 Simple game for entering into the market.

to compete with a monopoly firm, Old Inc. The monopoly firm does not wish the latest firm to enter the market because a new entry will decrease the incumbent firm's benefits. Therefore, Old Inc. threatens New Inc. with a price war if New Inc. enters the market. Figure 1.2 shows the extensive form of the market entry game. If the payoffs shown in Figure 1.2 are known to all players, the payoffs are 'common knowledge' to all players. If the game structure, including payoffs, is common knowledge, it is called a game of 'complete data.' The game tree indicates (1) New Inc. chooses to enter the market or not, and then Old Inc. decides to start a price war or not, and (2) the payoff of each decision combination.

A possible game prediction is that Old Inc. can use the approach: play 'start a price war' if New Inc. plays 'enter,' so Old Inc. can threaten New Inc. to stop them entering the market. As a result, it may seem that 'stay out' and 'start a price war if enter' are conditions satisfying the NE concept. Nevertheless, as shown in Figure 1.2, the threat to start a price war is not credible because Old Inc. will not create a price war if New Inc. does enter; instead, Old Inc. will maximize the payoff via playing 'no price war' after New Inc. enters the market unless Old Inc. is behaving irrationally. New Inc. knows the threat's lack of credibility and, therefore, will maximize the payoff by playing 'enter.' Thus, the price war game's NE is (Enter, No price war), an Eq that does not rely on the player to carry out a great threat. In a dynamic game, the Eq solution is a subgame perfect NE that satisfies sequential rationality by maximizing each player's payoffs in the subgames backward recursively (Gibbons, 1992).

A game of incomplete data or asymmetric information is also called a Bayesian game 'because it involves the use of Bayes' rules to solve the Eq. Therefore, static games of incomplete data are also called static Bayesian games. The core problem of incomplete data is the existence of 'private information' known only to notable players instead of all players. This private information generally refers to the payoff functions of the players. The players with private data are called informed players.

Similarly, the players who are uncertain of the other players' payoff functions are uninformed. The Eq of a Bayesian game is called 'Bayesian Nash equilibrium.' We won't be using this type of game in the models presented in the chapter.

After the introductions of incomplete data's static games, no additional explanation is required on a dynamic game of incomplete data. However, this class of games' possible equilibriums are much more complicated than other games yet still closely related. In this class of games, asymmetric data on payoff functions will be converted to asymmetric information on players' 'types,' where different players have different payoff functions. This class of game's central concern is to resolve the asymmetric information problem by differentiating the 'type.' Since many 3Ps games fall into this category, it is critical to fully understand their characteristics and the game's Eq concept: perfect Bayesian equilibrium (PBE).

In dynamic games concerning complete data, the subgame-perfect NE has to rule out incredible threats and time-inconsistent promises. However, in contrast to subgame-perfect NEs, the PBEs for dynamic games of incomplete data cannot be obtained through backward induction. PBEs need to be checked back and forth circularly. The primary steps to solve for PBEs are to:

(1) Find possible candidates for PBEs.
(2) Check each candidate or set of candidates for the satisfaction of PBEs in satisfactory solutions.

Two fundamental kinds of Eqs are often checked for possible conditions. The first is the 'pooling equilibrium,' under which different kinds of informed players act indifferently, and the uninformed player cannot differentiate the players' types according to their decisions. The second kind is the 'separating equilibrium,' under which different kinds of informed players act differently and, thus, the uninformed player can differentiate the kinds of informed players. For detailed explanations of why these solutions shall be satisfied, one may refer to Gibbons (1992) and Fudenberg and Tirole (1991).

Spence (1973) was the first to show how 'signaling' could solve the asymmetric information problem. He modeled that nature determines the laborer's productivity ability: high (H) or low (L), via probability p of being type H. Here we name the high-ability laborers as 'laborer H' and low-ability laborers as 'laborer L.' Second, the laborer learns his ability and chooses his education level as a signal: high education or low education. For simplicity, it is supposed that it costs \$0 for both types of laborers to achieve common knowledge and costs C_H and C_L to get a high education level for laborer H and laborer L. Third, the firm observes the laborer's education level. It offers the wage level, choosing high wages or low wages.

The two basic possible Eqs. to be checked are:

a) Pooling equilibrium:

Intuitively, if the low-ability laborer can achieve a high level of education as quickly as the high-ability laborer, the low-ability laborer would want to receive a high education level to convince the firm that he is a high-ability laborer. In this case, the firm cannot believe the laborer's signal, and therefore, will not offer a high wage level to the laborer with a high education level.

b) Separating equilibrium:

The solutions for the signal to be useful or for the condition to be a separating equilibrium shall be that it is not in the low-ability laborer's interest to imitate the high-ability laborer. In this case, the firm will believe the signal regarding productivity ability and offer wages accordingly. These conditions can be expressed mathematically. The conceptual intuitions of these conditions are as follows. First, considering the firm's wage offer decisions, the difference between the high-wage offer and low-wage offer must be large enough to compensate laborer H's extra education cost, but not large enough to compensate laborer L's high education cost. If the difference in wage offers is too large such that laborer L is willing to incur extra education cost, the signal will become ineffective. Second, considering the costs to different types of laborers, laborer H's signaling cost, C_H, must be less than laborer L's signaling cost, C_L, which is usually right in the job market signaling game if we believe that laborer H is smarter than laborer L. Cho and Krep (1987) further showed that the separating equilibrium would be the only possible equilibrium. Detailed discussion regarding the job market signaling game can be found in Spence (1973, 1974) and Cho and Krep (1987).

To build theories through game theory modeling, a game theory model will be developed to abstract the problem of concern properly. In this step, appropriate assumptions have to be made to simplify the problem to focus on a few critical components. In addition to game theory knowledge, this model setup process needs sufficient domain knowledge for the problem. A thorough literature review or case studies typically will provide a better and more precise understanding of the concerned situation and related problems.

The second step is to solve all possible or specific Eqs of the game model. The number of possible Eqs and the Eq solutions' complexity depend on the game model's complexity and the number of variables related to payoff functions.

The last step is to link the Eq conditions to the issues of the problem. If the equilibrium solutions are complicated, recognizing possible contextual or contingency variables will narrow the feasible solution space and provide more insights into the situation. Once the logic among different variable configurations and possible equilibriums is established, theories concerning the problem can be developed.

1.2 Issues related to unbalanced benefit structures in 3Ps

In 3Ps, the sources of the bidders' investment returns are the returns of equity investments in the concessionaire and the construction and operation contracts since the bidders would often act as the *major* contractors for construction and operation. Therefore, the bidders, being the controlling shareholders, will maximize the combined pool of benefit components' overall value. In other words, the bidders' benefit structure is inconsistent with that of concession firms' passive shareholders.

The benefit structure of a 3Ps investment can be better explained by the 3Ps business model illustrated in Figure 1.3, which shows that 3Ps investment returns include equity returns, construction contract returns, and operation contract returns. We will call the three benefit components together with the 3Ps 'benefit pool.' From the bidder's perspective, a 3Ps investment's benefits are the benefit pool's overall returns. Additionally, how these components are pooled in terms of their relative proportion will have major influences in determining the benefit pool's returns, the bidders' investment and potential opportunism, and the government and bidders' interactions.

- **First component: equity returns.** The benefit pool's first component is the equity returns in a 3Ps firm, illustrated as equity value minus equity investment, denoted as $E - I$. In 3Ps, following the project finance practice, the

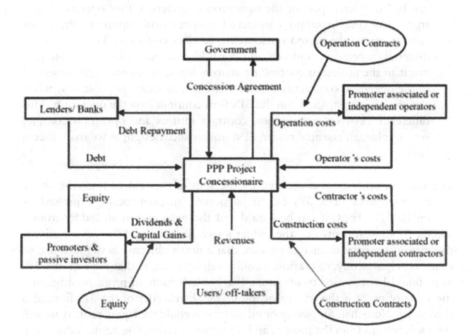

Figure 1.3 Bidders' benefit structure and benefit pool in 3Ps projects.

bidders will become one of the 3Ps firm's major shareholders, called 'controlling shareholders' in this chapter. The equity invested via non-bidders will be considered 'passive equity,' owned through 'passive shareholders.' Unlike the passive shareholders, like insurance companies, who mainly focus on equity investment returns, the bidders, being the controlling shareholders, will maximize the combined pool of benefit components' overall value. In other paths, the equity returns are not the only benefits sought by the bidders in a 3Ps project.

- *Second component: construction contract returns.* The construction contract returns refer to the benefits of bidders being the construction contractors of a 3Ps project. These construction-related returns are denoted as P_C, where subscript C represents construction contracts. As Walker and Smith (1995) observed, since most construction firms are thinly capitalized and rely heavily on short-term debt financing for their capital needs, they are generally reluctant to invest their limited and expensive capital in 3Ps equity and largely focus on construction contracts. This is especially true when the concession time is long and the returns from equity are slow.

- *Third component: operation contract returns.* The third component in the benefit pool is the returns from the bidder's operation contracts, shown as P_O, where O represents operation contracts. The operation contracts refer to the contracts for the daily operation and regular maintenance after the project is completed or the operation commences. For instance, the insurance policies for the facility properties, firm employees, or operation liability can be considered part of the operation contracts. Other operation contracts may contain supply contracts for operational inputs, contracts for regular maintenance, and outsourcing services contracts. Those who can undertake operation contracts and consider these contracts profitable may invest in the project as controlling shareholders. An essential characteristic of the operation contract returns is that the returns are possible only when the project is completed and the 3Ps firm continues via the operation. This difference between construction contract returns and operation contract returns plays an essential role in 3Ps' governance design, as we shall discuss later.

The value of profit pool is denoted by the sum of the returns from the above three components of a 3Ps benefit structure, mathematically expressed as $E - I + P_C + P_O$. Thus, it can be considered the net value to an bidder from a specific 3Ps investment, i.e., the bidder's overall benefits. Whereas traditional theories in corporate finance emphasize that a firm's objective is to maximize its value, or, equivalently, to maximize equity value, we argue that the aim of a 3Ps firm should be supposed to maximize the bidder's value of profit pool because the value of profit is the returns of the major shareholder of the 3Ps firm, also the bidder. Note that, for passive or minority shareholders, equity returns are still their sole returns from the project, and, therefore, maximizing equity value is the objective of their investment. We argue that the 3Ps bidder's deviation from the traditional objective in a non-3Ps firm creates serious problems, especially opportunisms when the benefit structure is unbalanced.

The skewed 3Ps utility structure defines an unbalanced utility structure in 3Ps to focus on short-term utility, particularly the PC's construction contract returns. The unstable benefit structure underlying 3Ps gives the controlling shareholders an incentive for opportunism. 3Ps' ownership structure further gives the controlling shareholders the capability to exploit private data in seeking appropriable rents from passive investors. The controlling shareholder may profit from manipulating the construction contract prices and clauses. As an outcome, the minority shareholders, subject to severe data asymmetry, will suffer from losses in equity returns. They control shareholder–passive shareholder conflicts that may seriously impair the project's financial situation and performance and lead to significant transaction costs in 3Ps.

In the cases where equity is allowed to be raised publicly before project completion, the passive shareholders will be in an even weaker position because of the severe information asymmetry in unfinished projects and the relief of the controlling shareholder's equity investment requirement. Additionally, since the use of project finance in 3Ps allows a low equity ratio, the danger of early public equity raisings based on an unbalanced project structure problem is further aggravated.

1.3 Issues of financial renegotiations/hold-up in 3Ps: dynamic game of complete data

Financial renegotiation and the related hold-up problems may happen when project cost, market demand, or other market conditions become significantly unfavorable and cause the bidder to renegotiate with the government for subsidies or rescue. The dilemma encountered by the government is that although financial renegotiation is not legitimate, the government is often tempted to approve the renegotiation because of the gigantic costs of project failure. Such inconsistency creates serious opportunism issues and associated transaction costs. Ho (2006) developed a game theory model that analyzed the strategic interactions concerning renegotiation/hold-up and proposed related procurement and management policies from renegotiation/hold-up. We shall present the model and argue about how the problem is modeled and its major policy implications.

The expectation that the government will renegotiate under project distress may cause opportunistic bidding, generally seen in construction practice. In opportunistic bidding, bidders, in their proposals, intentionally understate the possible risks engaged or overstate the project profitability to outperform other bidders. In their pilot research on opportunistic bidding, Ho and Liu (2004) show that if a builder can make an effective construction claim, the builder will have a stimulant to bid opportunistically. Similarly, if a request for renegotiation is always granted, bidders would have a stimulant to bid optimistically or aggressively to win the project. An overly optimistic proposal can have a higher chance of winning given that many crucial and bidder-specific project data in the bid proposal can be complicated to verify, and the government tends to favor those proposals with attractive financial forecasts. Therefore, if the bidders have an ex-ante expectation of ex-post renegotiation, they will have a stimulant to bid opportunistically. As this logic under governments saving subsidies due to renegotiation and the project's early failure due to opportunism is not straightforward, the importance of the financial renegotiation problem is underemphasized.

In his repossession game instance, Rasmusen (2001) shows that if renegotiation is expected, the agent may choose inefficient actions that will decrease overall or social efficiency but enhance the agent's payoff. In 3Ps, after signing the concession, moral hazard issues may also happen if renegotiation is expected. As bidders are frequently the major contractors of 3Ps projects, they may not be concerned about cost overruns because they may profit from such overspending. Additionally, bidders may not be concerned by the operation efficiency either during the operation period.

1.4 Modeling of financial renegotiation/hold-up and the equilibriums

The behavioral dynamics of the renegotiation, or government rescue, play a central role in 3Ps management given that data asymmetry generally exists. Here, game theory is applied to analyze how the government will respond to the bidder's request for renegotiation and the effect of such renegotiation on 3Ps.

The game theory framework for analyzing a 3Ps investment shown in Figure 1.4 is a dynamic game explained in an extensive form. Suppose a 3Ps contract does not determine any government rescue or subsidy in the encounter with a financial crisis. The law prohibits the government from bailing out the 3Ps project by providing a debit guarantee or extending the concession period.

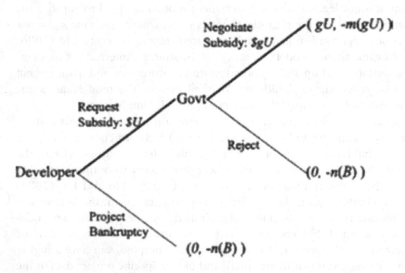

Figure 1.4 Renegotiation game's Eq. structure.

Suppose also that the government is not encouraged to rescue a project without compelling and justifiable reasons. Cost overrun or operation losses caused via inefficient management or average business risk are not considered reasons for government rescue. In contrast, adverse events caused by unexpected or unusual equipment/material price escalation may be justified more easily. It shall be reasonable to assume that if the government grants a subsidy to a project based on unjustifiable reasons, the government may suffer a loss of public trust or suspicion of corruption.

The dynamic game, shown in Figure 1.4, starts from adverse situations where it is in the bidder's (denoted by D in the game tree) or lending bank's best interests to bankrupt the project if the government (represented by G) does not rescue the project. Alternatively, the bidder can also need government rescue and subsidize the amount of $\$U$, even though the contract clause does not determine any possible future rescue from the government. Here U is denoted as the present value of the net financial viability change and is regarded as the maximum possible requested subsidy. Note that U is not the actual subsidy number. The existing subsidy is determined in the renegotiation process discussed later.

On the other hand, if a 3Ps project is bankrupted, the payoff of government is $-b(G+\tau)$, the *political cost due to project retendering*, where G is the minimum government funds necessary for the financial redesigning of a project when a project is bankrupted and retendered, and τ is the opportunity cost for remodeling bidders, which may include the retendering cost and the cost of discontinuity due to the bankruptcy and rendering process. Suppose that for a 3Ps project to proceed further onto the procurement stage, the project must have been shown to provide facilities or services that can be justified economically. In this game, it is supposed that retendering is desired by the government, as often observed in practice if a project is going bankrupt.

Alternatively, as Figure 1.4 represents, the bidder can negotiate a subsidy starting with the maximum amount of $\$U$. The subsidy can be in various forms, such as a debt guarantee or concession period extension. Generally, the bank will not provide extra capital without a government debt guarantee or other subsidies. Because the debt guarantee is a liability to the government, and an asset to the bidder, a debt guarantee is equivalent to a government subsidy. Other forms of subsidy may include extending the concession period, more tax exemption for a certain number of years, or an extra loan or equity investment directly from the government.

After the bidder requests the subsidy, the game proceeds, as shown in Figure 1.4, to its subgame: 'negotiate subsidy' or 'reject.' If the government rejects the bidder's request, the project will be bankrupted and retendered. The payoff for both parties will be $(0, -b(G+\tau))$. If the government decides to negotiate a subsidy, expressed via the *rescuing subsidy ratio g*, a ratio between 0 and 1, the payoff to the bidder and the government will be $(gU, -b(gU)-r(gU))$, respectively, where $-b(gU)-r(gU)$ is the *political cost due to the rescuing subsidy to a private party*, containing two functions b and r, as we shall argue later. Rescue a 3Ps project and providing a rescuing subsidy to the original 3Ps firm could provoke severe criticism of the government. If the government lacks compelling reasons for the subsidy, the criticism will cause significant *political* costs related to the magnitude of the subsidy. The differences between the two functions will be discussed in detail later. Here 'g' is not a constant and is used to model the process of 'offer' and 'counter-offer.' More details on negotiation modeling using g can be found in Ho and Liu (2004).

- ### Political cost of bailing out

Suppose the government negotiates the subsidy with the existing bidder and rescues the project. In that case, the function of the political cost to government is modeled here as $b(gU) + r(gU)$, is the function of the political cost and $r(\bullet)$ the political cost of over-subsidization. The mathematical modeling of the political cost of bailing out is according to the fundamental concept that resources are scarce. If the government has unlimited funds to spend, there would be no political cost for bailing out a project. Since the government only has a limited budget, there will be a political cost should the additional funds be needed for the project. The more government funds are required, the higher the political cost is. Here $b(gU)$ measures the political cost caused by the budget burden from subsidies and is considered the 'basic' political cost. As shown in Figure 1.5, guarantee of the political cost should *increase* the amount of subsidy gU.

Moreover, on top of the essential political cost, it is supposed that further political costs would be incurred to reflect a severe resource misallocation for the subsidy exceeding a justifiable amount. To define the function, $r(\bullet)$, we shall first define J, the amount of the subsidy that can be justified without criticism of over-subsidization; in the model, 'J' is termed the 'justifiable subsidy,' which is considered by the public an eligible claim for subsidy. 'J' can be measured by imagining the amount of '*claim*' that could be granted to the bidder had the case gone to court. For instance, the damages due to force majeure might be considered justifiable. If the subsidy is less than the justifiable claim, the government will not be blamed for over-subsidization.

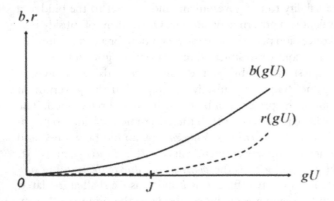

Figure 1.5 Costs of guarantee.

Therefore, as illustrated in Figure 1.5, $r(\bullet)$, it can be defined by that $r(gU)$ is zero when $gU \leq J$ and that $r(gU)$ becomes an enhancing function when the subsidy is more generous than J, meaning that the government will be criticized for over-subsidization or suspected of corruption and will suffer further political cost in addition to the essential political cost. The overall political cost of bailing out, $b(gU) + r(gU)$, as shown by the kinked curve in Figure 1.6, is obtained by adding the two components in Figure 1.5.

- *Political cost of project bankruptcy*

Considering the lending bank can effectively monitor the project's financial status, it may be inferred at the time of bankruptcy that the project's overall value is less than or close to the estimated total outstanding debt. As a result, it is unwise for the bank to continue providing additional capital under near-bankruptcy conditions. Thus, the lending bank will deny further capital requests, even when such capital is still inside the project's original loan contract.

When a project is bankrupted, it will be regarded as 'sold' to the government and retendered to another private bidder, assuming that the project is still worth completing. The government may want to regain control of the project for retendering because a 3Ps contract is generally associated with public services and cannot be transferred directly to a new bidder without a new concession. From this point of view for the government, bankruptcy is equivalent to a costly replacement of the bidder. Because of the use of project fiscal solutions in 3Ps, the project to be retendered by the government will still be mainly financed via debt. As an outcome, when a project is bankrupted, the number of budgeting burdens can be modeled $G + \tau$. Following the function's definition, $b(\bullet)$, project bankruptcy's political cost can be modeled $b(G + \tau)$, as shown in Figure 1.6.

As noted previously, the fiscal renegotiation game tree derived above will be solved backward recursively, and its NE conditions will be obtained. Since the

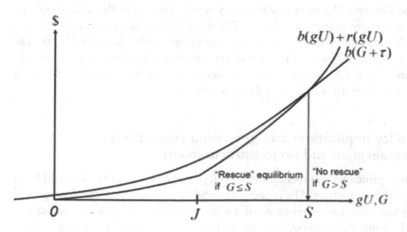

Figure 1.6 Solutions for 'rescue' Eq and 'no rescue' Eq.

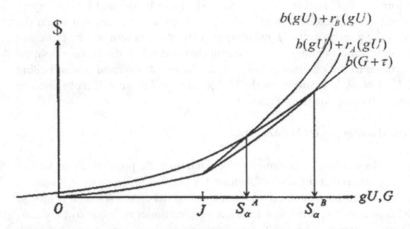

Figure 1.7 Effects of the change $r(\bullet)$ on the Eqs.

values for the game's payoff matrix variables are undetermined, the payoff comparison and maximization cannot be solved for a unique solution. However, the conditions for possible NEs of the game can be analyzed. There are three candidates for the NEs: (1) the bidder will 'request a subsidy,' and the government will 'negotiate a subsidy,' (2) the bidder will 'request a subsidy,' and the government will 'reject,' and (3) the bidder will select 'project bankruptcy.' The first Eq should be called 'rescue equilibrium,' and the second and third Eqs will be called 'no rescue equilibriums.'

Ho (2006) solved for the solutions related to 'rescue equilibrium' and showed that it is impossible to rule out the 'no rescue equilibriums.' Figure 1.6 illustrates that when G is less than S, with the curves' intersections $b(gU) + r(gU)$ and $b(G + \tau)$, the rescue equilibrium will be obtained. Furthermore, the bidders will expect this condition for rescue equilibrium, inducing opportunistic behaviors from the bidders. The most crucial policy implication from the analysis of Ho (2006) is that government policies on 3Ps should try to decrease the magnitude of S to decrease the possibility of opportunism from the bidders. For instance, as shown in Figure 1.7, when the slope of the function, $r(\bullet)$, becomes steeper due to specific policies, S's magnitude will be decreased significantly. Similarly, S can also be effectively decreased when J or τ is reduced.

1.5 Policy implications and governing regulations for management and project procurement

Governing principles and management policy implications can be obtained from the game theory analyses. The proposed model does not provide the strategies to quantify the game parameters; this pilot research concentrates on the characteristics of the game parameters/functions and their relationship. The concentration will be on which strategies can decrease the renegotiation problem and enhance the administration in 3Ps. Possible governing principles and administration policies for 3Ps projects are given as follows.

Principle 1: be well prepared for renegotiation issues, as it is impossible to rule out the possibility of renegotiation and the 'rescue' Eq.

Since it is impossible to rule out the 'rescue' Eq, the government should be well prepared for the opportunism problems induced by the *ex-ante* expectation of renegotiation, as argued previously. Policy implications from this principle include:

- In project procurement, the government should identify the possibility of opportunism problems and cannot consider the bidder's fiscal plan as a binding, credible contract.
- The government could devise a mechanism to persuade bidders to reveal more private information. For instance, the government can establish a formal policy to disqualify an bidder if they are shown to have a history of behaving opportunistically.

Principle 2: although renegotiation is always possible, the probability of reaching 'rescue' Eq shall be minimized and reduced by strategies that increase the political cost of over-subsidization $r(\bullet)$, and decrease the bidder-replacement cost τ, and the justifiable subsidy, J.

One way to reduce the opportunism issues is to minimize the probability of 'rescue' equilibrium and the bidder's expectation of the probability. Policy implications of this principle may include:

- Laws may regulate the renegotiation and negotiated subsidy, and such laws will enhance $r(\bullet)$, when the subsidy is not justifiable.
- Adequate monitoring or an 'early warning' system can give the government enough lead time to prepare to replace an bidder with minimal effect, and hence, reduce τ.
- To decrease J, the government should pay attention to the contract's quality in terms of content and implementation, e.g., the scope, risk allocation, documentation, and contract administration process.

Principle 3: the government shall specify a fair, justifiable subsidy, J, which corresponds to the bidder's responsibilities and allocated risks determined in the contract.

Holliday et al. (1991) discuss that they are often bidder-led because of 3Ps projects' scale and complexity. It is tough to recognize a clear client–contractor connection. The 'bidder-led' phenomenon implies data asymmetry and an opportunism issue in 3Ps projects where the bidder may hide data and have an incentive to behave opportunistically. Policy implications may include:

- The government can separate the bidder from the builder/contractor in a 3Ps project to have a more transparent client–contractor connection.

- The government can assign third-party experts to serve on the project company's board to ensure proper monitoring and accurate information collection.
- Risk assignment among the concessionaire and government should be made explicit in the agreement. This could assist in specifying a fair J in the future.
- The government shall carefully specify when they may intercede and what they shall do. Via temporarily taking over a project, the government should focus on having more information regarding J and G and reducing τ through gaining longer lead time to prepare for the retendering.

1.6 Bidders' 3Ps approaches for signaling: dynamic games of incomplete data

Signaling games are the most widely applied class of dynamic games of incomplete data. The first move of such a game is initiated via the player, often called 'Sender.' The Sender has private data considering his type, e.g., low productivity ability or high productivity ability. In the 3Ps signaling game, the bidder is the Sender of signals, and the government is the receiver. The main opinion is that effective communication can occur if one player is willing to send a signal that would be too expensive for the other kinds of player to send (Gibbons, 1992). Thus, it is crucial to know which signals to reach the separating equilibrium in the 3Ps signaling game. The signaling game analysis is expected to provide both the bidders and the government with deeper insights into 3Ps strategies.

In a 3Ps project, the bidder will send signals to the government to convince the government of the bidder's type. From the perspective of an unbalanced benefit structure problem, the bidder's types can be categorized into 'long-term benefit-oriented' (LT type) and 'short-term benefit-oriented' (ST type), which can be specified via the relative magnitudes of different components in the benefit structure. Instances of signals may include the project cost's equity level, the financial projections, the self-exclusion of being a construction contractor, etc. One can also view the proposal itself as a collection of signals. After receiving the 'signals' or proposals, the government's actions would be the proposal evaluation scores that lead to project awarding.

1.7 Separating equilibriums of the 3Ps signaling game and equity ratio as a signal

The following analysis will be at the conceptual level, instead of technical details, so that the readers won't be distracted by the technical difficulties. In a 3Ps project, the bidder's payoffs are the value of profit pool, as discussed earlier. The value of profit pool and its three components' relative magnitudes depend on the bidder's type, unknown to the government. It is natural for the government to favor a type of long-term benefit-oriented bidder over a short-term benefit-oriented bidder. In a signaling game, the long-term benefit-oriented bidder may send *costly signals* to signify the bidder's type. The costs of the signals are mostly reflected in the effects, mostly negative, on the value of profit pool. For the signal to effectively reach the separating equilibrium, the long-term benefit-oriented bidder's signal's costs must be significantly lower than the short-term benefit-oriented bidder's costs for sending the same signal. If a signal's costs must render the value of profit pool negative, the bidder will not send such a signal.

As argued previously, it is supposed that the government's major concern is the project's financial viability, reflected via the returns to the shareholders, specifically- $E - I$, the first component of the value of profit pool. However, the government's concern for the value $E - I$ is inconsistent with the bidder's concern for the value of profit pool; as discussed previously, with a highly unbalanced 3Ps benefit structure, the bidder may have a loss on equity investment, i.e., negative $E - I$, but still enjoy a high overall value of profit pool due to the significantly higher P_C. Therefore, in a signaling game where the government is unsure about the bidder's type, the long-term benefit-oriented bidder's goal is to send the signals that convince the government that they have positive and reasonably good returns on equity investments. For example, some 3Ps bidders use a 'high equity ratio' as a signal for positive $E - I$. If the project is not fiscally viable, the high equity ratio will yield tremendous losses in equity. However, using a high equity ratio also raises the 3Ps investment costs since the required return rate/capital costs for equity are much higher than that of debts or bank loans. Here the high equity ratio may constitute a separating Eq because the short-term benefit-oriented bidder usually cannot send the same signals. The use of a high equity ratio is too costly for the short-term benefit-oriented bidder because, without the compensation from the construction contract returns, the bidder may negatively value profit due to the losses from equity investment.

Note that a signal may be useful only under the game Eq, that is, if the government can differentiate the signal's quality or costs. If the government does not recognize the signal's cost difference, then the long-term benefit-oriented bidder may not be willing to convey such a costly signal.

1.8 Possible signals and effectiveness

Previously, it was concluded that the bidder's high equity ratio could be a useful signal. Further analysis of the effectiveness of other potential signals sent by bidders can be evaluated. Here, we will argue the effectiveness of some other famous signals sent by bidders. Note that the conclusions on the signal effectiveness can be different based on different contexts.

It is reasonable that the government prefers those proposals that offer lower project costs. However, since 3Ps projects are not procurement via traditional design-bid-build or design-build schemes, the proposed project costs are not paid via the government and not binding; there are virtually no subsequent costs/ losses for the bidder for proposing understated, untruthful project costs. As an outcome, low project costs should not be a useful signal for separating equilibrium. In the off-Eq situation where the government does not realize such a signal's ineffectiveness, if the low project costs are one of the government's essential evaluation criteria, the short-term benefit-oriented bidder may have stimulants to imitate the low-cost bidders to enhance the proposal's winning probability. Consequently, if the government takes low project costs as a good signal, the bid proposals' cost information will be distorted.

The proposal's future operating cash flow or toll revenue projection is also one of the project's most critical financial figures because its financial viability mainly depends on cash inflows. However, they may also be the most unreliable figures in the project proposal. Like the project cost signal, since the projection of project cash inflows is not binding and the costs of reporting an overly optimistic projection are minimal, this signal shall be ineffective, not to mention the difficulty of approximating the long-term revenues concession period. Unfortunately, inexperienced governments often take these cash inflow projections as critical signals without evaluating their credibility or quality. When the wrong signals are taken, the information will be distorted, and the government will choose the wrong bidder.

The concession period and benefit-/risk-sharing scheme shown in the proposal are almost binding provisions, although the concession contract has not been signed. Therefore, the concession period and benefit-/risk-sharing scheme will impact the value of profit pool for different bidders. These two signals might be more effective in general.

The government's subsidies may be in the form of a government debt guarantee, operating revenue guarantee, or direct subsidies for construction costs. These guarantees are typically provided only when the project's future revenue is not high enough or too uncertain. The project cannot be fiscally viable without the government guarantee. In this case, some might require an exact amount of government guarantees. However, the government generally prefers that the bidder does not need any government subsidies or guarantees. However, it is difficult to enforce such a promise that appeared in an offer, especially when the government cannot tolerate project failures and does not have clear policies considering post-awarding negotiation or financial renegotiation. On the other hand, if governments are subject to be easily held up, requesting no proposal subsidies shall not be considered a good signal.

Alternatively, we discussed that the government should thoroughly analyze the 3Ps project's financial feasibility from both the bidder's and shareholder's perspectives and decide whether subsidies are necessary before requesting proposals. If it is decided that the subsidies may be necessary, the government shall form and announce the subsidy policies upon the invitations for proposals. Simultaneously, the government should do its best to prohibit or discourage any post-awarding requests for the subsidies if the bidder did not ask for the proposal's subsidies. In such a context, requesting no subsidies might become a sufficient, credible signal.

The letter of commitment implies that the project will be financed with a reasonable interest rate. Because the resources and costs of debt play an essential role in 3Ps projects' success, a proposal without a lending bank's commitment may mean the bidder will request a debt guarantee or interest subsidization after the project is awarded. Therefore, it may be easier for the long-term benefit-oriented bidder to gain a letter of commitment from the banks. As a result, the lending banks' written commitment is generally regarded a good signal in practice if the lending banks are reputable banks with 3Ps experience. However, since the banks are generally not bound by the commitment and can refuse to provide loans in the concession/contract negotiation stage, some banks may not be so severe or conservative in their commitments in order to retain the possible business options. In this case, the effectiveness of this signal may not be as good as we have expected.

By definition, the long-term benefit-oriented bidder's major benefits are mainly from equity returns $E - I$ and operation connected with contract returns, P_O, rather than construction contract returns, P_C; thus, it may be less costly for the long-term benefit-oriented bidder to exclude himself from becoming the future construction contractor. This signal shall be very effective. However, the signal may cause additional costs to the 3Ps firm/project, rather than an bidder, because of the sacrifice of the profits of better design integration through contractors actively participating in the early stage. The additional costs may render the proposal an unfavorable solution in competitive bidding. Moreover, even for the long-term benefit-oriented bidder, many 3Ps projects cannot be fiscally feasible without the extra returns from partially undertaking the construction contracts.

2 Corruption and its impact on government policies and evaluating Public–Private Partnership

2.1 Government 3Ps approaches for evaluating: dynamic games of incomplete data

Alternatively, governments can play evaluating[1] games by becoming the first mover and setting criteria for evaluating out unwanted developers. 'Evaluating' was used to refer to the market process studied in Rothchild and Stiglitz (1976), in which the information problem in the insurance model was similar to that in Spence's (1973) job market model as discussed previously in the introduction of game theory. In the evaluating game, the Receiver initiates the move, and the Sender responds. An effective evaluating scheme design can induce the Sender to reveal private information or types. For example, in the job market analysis, the firm may make the first move by specifying a menu of contracts regarding the education levels and wage offers. The workers will then respond by selecting their preferred contracts. In the job market, a useful evaluating menu of contracts would induce worker H to choose the contract with a high education level and a high stipend, and worker L to select a low education level and a low stipendcontract.

On the other hand, an ineffective evaluating scheme cannot induce the workers to reveal their types. For instance, if the wage level difference is too large, it may be optimal for worker L to imitate worker H, invest more in education, and obtain a high stipend. If the stipendlevel difference is too small, it may be optimal for worker H to select low education, thus eliminating the need to incur high education costs.

The use of evaluating criteria can be regarded as governments' strategies for the implementation of 3Ps. The central issues and the game equilibrium concepts are very similar to those in signaling games. A useful evaluating criterion must be that only those of the desired type can satisfy the requirement with reasonable costs. We argue that since 3Ps projects are usually initiated by governments, who often have specific goals to achieve, playing evaluating games seems more sensible and practical for governments than waiting for the developer or bidders to guess what the government wants to send out weak signals.

DOI: 10.4324/9781003177258-2

In this chapter, we describe a 3Ps as a contract among the government and a private-sector company under which:

- Private company finances build and operate some element of a public service.
- The private sector gets paid over several years, either through charges paid by users, or with payments from the public authority, or a mixture of both.

3Ps are now being developed worldwide via global institutions and consultants. Development banks, national governments, the European Union, and donor agencies provide subsidized public fiscal specifically for 3Ps. Nations subject to International Monetary Fund regimes, and other developing countries, are being subjected to political pressures and marketing campaigns.

But experience over the last 15 years shows that 3Ps are an expensive and inefficient way of financially supporting infrastructure and divert government spending away from other public services. They conceal public borrowing while providing long-run state guarantees for advantages to private companies.

This chapter looks at the scale of 3Ps and the institutions improving them, the lessons of experience with 3Ps, and a procedure for systematic analyses of 3Ps against public party alternatives. It also sets out some ways of challenging 3Ps policies and programs and offers advice to pension funds regarding investing in 3Ps.

The fiscal crisis of 2008 onwards brought about renewed interest in 3Ps in both developed and improving nations. Encountering constraints on public resources and fiscal space while recognizing the crucial need for investment in infrastructure to assist their economies to grow, governments are increasingly turning to the private section as an additional alternative resource of funding to meet the funding gap. While recent attention has been concentrated on fiscal risk, governments look to the private sector for other reasons:

- Exploring Public–Private Partnerships as a way of illustrating private-sector technology and innovation in providing better public services through developed operational efficiency
- Incentivizing the private party to deliver projects on time and within budget
- Imposing budgetary certainty via setting present and future costs of infrastructure projects over time
- Utilizing Public–Private Partnerships as a way of improving local private-sector abilities through joint ventures via large international firms, as well as subcontracting opportunities for local businesses in areas like civil works, electrical works, facilities administration, security services, cleaning services, maintenance services
- Using Public–Private Partnerships as a way of gradually exposing state-owned enterprises and government to enhancing levels of private-sector engagement (mainly foreign) and structuring 3Ps in a way to ensure the transfer of skills to national champions that can run their operations professionally and eventually export their competencies by bidding for projects/joint ventures

- Creating a model in the economy via making a nation more competitive in terms of its facilitating infrastructure as well as giving a boost to its firms and industry according to the infrastructure improvement (like support services, construction, equipment)
- Supplementing limited public-sector capacities to meet the growing demand for infrastructure development
- Extracting long-run value for money through appropriate risk transfer to the private party over the life of the project – from design/construction to operations/maintenance

As argued previously, the Developer's maximization rationale should be applied to the investment's value of profit pool instead of the 3Ps firm's equity value. As a result, the Developer will make decisions according to the value of profit pool maximization rationale. However, if the government imposes specific 3Ps criteria or policies for evaluating, the Developer's optimal decision will be affected by the policy. For example, if the government is constrained by a law or policy that forbids any forms of subsidy for rescuing distressed projects, 'no rescue' will be the equilibrium of the fiscal renegotiation game and, expecting that it will be more challenging to hold up the government, the Developer will then be less likely to bid aggressively.

Smith (1999) maintained that 'by the start of the twentieth century, the connection among the private sector and government in infrastructure procurement had begun to reach some kind of maturity.' He summarized that the relationship had made it possible for the government to play more roles today than before, including regulator, customer, facilitator, investor, planner, protector of the public interest, defender of the realm, guarantor, agent of economic change, and supporter of export trade, etc. Depending on differences in countries and economic and political environments, the government's roles are different. As a result, the public's different government roles will result in various policies, and multiple policies will have critical impacts on the government's attitudes or strategies toward 3Ps. For example, governments in emerging countries that mainly act as planners and agents of economic change may not have the policies against subsidizing a distressed 3Ps project, since the government's primary role is to provide the infrastructure and boost the economy. The fairness concern may not be as important as in developed countries. In contrast, a government that mainly acts as a protector of public interest may have tighter regulations on the 3Ps project's tendering and contract management to prevent corruption, over-subsidization of a project, or a project's unreasonably high returns.

Under a competitive scheme, a 3Ps project will usually be given to the proposal winning the highest overall evaluation scores. Thus, the evaluation criteria for assigning scores to a request will be crucial to the Developer's improvement and bidding strategies. In 3Ps, government evaluating strategies may be transformed into evaluation criteria. From a separating equilibrium perspective, useful evaluation criteria will have the developers self-select into different needs according to their characteristics so that only the desired developers will be selected or stand out.

2.2 Evaluation of some government 3Ps evaluating approaches

This section will discuss the effectiveness of some famous or possible government evaluating strategies or developer evaluation criteria. Since the evaluating games are almost the same as the signaling games except for the moving sequence, the signals whose effectiveness was discussed previously can be applied to the 3Ps evaluating when these signals are used as evaluating criteria.

According to Tiong (1996), among those critical success factors in winning a 3Ps project, 'fiscal viability' was recognized as one of the essential elements. Tiong (1996) listed some typical government evaluation criteria for a 3Ps project's fiscal package, including high equity level, low construction cost, acceptable tolls/tariff levels, and short concession period. In practice, governments typically will weigh the project's fiscal viability heavily in evaluating a proposal. The weighting of fiscal feasibility can range from 20% to 50%. As discussed in signaling games, the problem in focusing on the fiscal package is that the Developer's 'true' estimated figures are often unknown to the government. The figures in the Developer's proposal may be manipulated to win the project. Therefore, overly emphasizing the fiscal plan will encourage the bidders to use optimistic estimations and favor aggressive bidders.

As Tiong (1995) stated, equity level or equity ratio requirement is commonly specified in request for proposals. In addition to the minimum equity ratio required, some request for proposals may further state that a higher equity level is preferred. Tiong (1995) argued that the rationales behind this criterion are (1) high equity will reduce the project's debt burden; (2) it signifies the Developer's faith in the project's viability; and (3) it may motivate the Developer to complete the project on time and budget. From the perspective of game theory, these rationales mean that high equity may either be a valid *signal* to signify the Developer's private information, such as the project's viability, or *screen out* unqualified developers. We may infer that a high equity ratio is usually an effective strategy because, for developers with a profit structure skewed to construction contract returns, it may be more challenging to counterbalance the loss due to equity investment when the equity ratio is high. Therefore, a high equity level requirement may screen out those developers who don't have a fiscally viable evaluating rationale.

Nevertheless, a high equity ratio as a criterion should be used with caution because the effectiveness is not guaranteed. The costs of this criterion to the developers or society are very high. First, suppose the profit structure is highly unbalanced and skewed to the construction contract returns. In that case, it is possible that the loss due to a certain level of higher equity ratio can still be compensated. Second, the equity ratio for evaluating purposes should exclude the equity from passive shareholders because passive shareholders' equity is not related to the Developer's value of profit pool. In other words, the high equity ratio should be directly contributed by the Developer. Third, using a high equity ratio will deviate from the spirit of the project financing arrangement in 3Ps and significantly increase capital cost and eventually be imposed back on the project users. It is tough practically for developers to finance a large-scale project with a high equity ratio. As a conclusion, the increased equity ratio criterion may discourage the participation of private parties.

As discussed earlier in the renegotiation model, under the 'no rescue' equilibrium, the government will not provide subsidies should critical adverse events occur. With explicit policy on limiting the government's rescuing subsidy, subsidies for recovering concession firms will be difficult to justify. Because of the awkwardness of holding up the government for renegotiation, the value of profit pool of the short-term benefit-oriented developer will be reduced significantly, and the short-term benefit-oriented developer will eventually be screened out before or during the tendering process.

Note that the *non-confusion* of the 3Ps policy is no less important than the policy itself. If the government does not specify the policy regarding the post-awarding renegotiation, there will be two consequences. First, prudent and responsible developers may assume that there will be no post-awarding subsidy and evaluate the investment accordingly. Second, aggressive developers will be better positioned in bid competition by factoring in that the 3Ps project will have a 'rescue' equilibrium through renegotiation. In the proposal selection, the government may favor the aggressive Developer's proposal since the proposal's figures based on the 'rescue' equilibrium will outperform others' suggestions. In this scenario, the government will fail to select a responsible developer. In contrast, if the government specifies the conditions of post-awarding subsidies, all developers will explicitly evaluate the subsidy's value on the same ground. In this case, the government will have a higher chance of awarding the project to a better developer.

2.3 A contingency view of 3Ps as a governance structure: when and why?

While 3Ps provide a new alternative for delivering infrastructures, many 3Ps failures have shown us that 3Ps, under certain situations, can be the wrong governance structure for providing infrastructures. However, few theories offer a systematic view on when 3Ps are or are not a good governance structure and, more importantly, why. Here we try to answer these questions via integrating the transaction cost economics and game theory view of 3Ps.

While the greater efficiency due to better pooling of resources and high-powered incentives is mostly emphasized in 3Ps, the effects of transaction costs embedded in Public–Private Partnerships are often understated. As we have observed in practice, the high transaction costs could render Public–Private Partnerships a cheap alternative for providing infrastructure. The definition of transaction costs here focuses on the costs due to opportunistic behaviors or the prevention of opportunistic behaviors. Different governance mechanisms represent different tradeoffs among advantages and transaction costs. Therefore, selecting from the different schemes as an option for infrastructure deliveries entails careful analysis on the comparative tradeoffs among transaction costs and benefits.

From the transaction cost economics perspective, there might be distinctive and substantial transaction costs embedded in 3Ps because of the potential opportunisms. These transaction costs include those that can be observably recognized and measured and hidden, not easily assessed. The hidden transaction costs may significantly undermine the expected benefits and sometimes cause disastrous impacts on society or project success. Specifically, we argue that there are two major sources of transaction costs in 3Ps, namely, the unbalanced profit structure problem and the hold-up induced renegotiation problem that were discussed in previous sections. The magnitude of the transaction costs thus depends on many factors associated with the two issues. Different types of projects in different environments or institutions will have significant extra transaction costs; therefore, 3Ps' contingency theory as a governance structure is desired.

Ho and Tsui (2009) suggested an algorithm, as depicted in Figure 2.1, for the strategic interactions between governments and developers in 3Ps. This model integrates the signaling view and the evaluating view of 3Ps. The government needs to establish evaluating policies that may decrease the incentives for and availability of opportunism. Similarly, the developers will make their investment

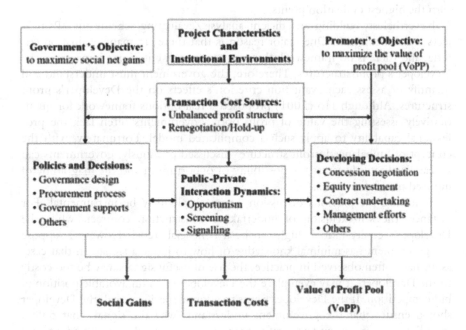

Figure 2.1 The framework of 3Ps.

decisions, such as concession negotiation, equity investment, and management efforts, etc., either as an answer to the government techniques or as signals of their types. All decisions made by the players are jointly determined by each player's objective and the interaction dynamics/game being played. Mainly, we assume that the government aims to maximize the overall social welfare (through which the government also maximizes the political benefits). The Developer's objective is to maximize the 3Ps profit pool's overall returns, i.e., value of profit pool. The degree of goal incongruence influences the interactions among the government and the developers.

In equilibrium, project characteristics and institutional environments related to information asymmetry, renegotiation, and other opportunism problems will affect the parties' interaction dynamics and decisions. We shall identify these project and institutional factors later for determining whether 3Ps are a good governance structure.

2.4 The difficulties of evaluating and signaling: problems and causes of off-equilibrium

In 3Ps, the government can freely specify any combination of evaluation criteria and their weights. If the government determines the requirements and consequences in the project's RFPs, the RFPs will become the menu of contracts in a evaluating game. An effective evaluating scheme may induce each Developer to *self-select* into the contract or evaluation scheme so that the best Developer will earn the highest evaluation points.

Nevertheless, building an efficient analysis/evaluating scheme for 3Ps projects is a difficult task. One major reason is that there are many evaluation criteria, and there is no framework that can assess each criterion's impacts on the Developer's profit structures. Therefore, the government must understand and quantify or assess each evaluation criterion's effects on the Developer's profit structures. Although Ho (2001) proposed a real options framework for quantitatively assessing the value of profit pool, governments often lack the professional capability to apply such a complicated model. Fortunately, with the concept of 'unbalanced profit structure' discussed previously, governments can at least qualitatively assess the possibility of opportunism and the magnitude of induced transaction costs.

Following the previous discussion, some signals may be instrumental. For instance, the self-exclusion of undertaking construction contracts while the Developer's equity ratio is high can be a beneficial signal. However, suppose the government has minimal knowledge of how to judge a signal. In that case, as we have often observed in practice, the use of useful signals may be too costly to the Developer. It may even place the Developer in an unfavorable position in bid competition. If the Developer wants to adopt a specific signal, the Developer should ensure that the government understands why the signals that only a 'good' developer can send are useful. Sometimes, the Developer who adopts and sends out proper signals should take the responsibility to educate the public and government on why the signals are useful. Unfortunately, this is a challenging task due to the conflict of interest and the government's lack of professional ability, and slow learning.

To complicate the problems, if the government is too naïve or too inexperienced in 3Ps, the transaction costs caused by opportunism will be high, much higher than the prices for preventing the opportunism through signaling and evaluating. 3Ps will become an inappropriate governance structure if the transaction costs are too high. For instance, given specific institutional environments and project characteristics, it is very likely that the evaluating strategies by the government will be either too stringent or too easy to satisfy and that the government will not recognize the signals sent by the developers/bidders as an indication of developers' types. In this solution, the results will be 'off-equilibrium,' and the probabilities of project failure will be high. When the project's loss is unendurable and unacceptable to the government, the government will be easily held up by the concessionaire after the concession has been granted. There are few choices for the government but to bail out the project or provide *ex-post* subsidies in the event of distress. Having foreseen the government's actions, the Developer would submit opportunistic bids *ex-ante* to win the concession and then appropriate excess profits *ex-post* from the government, at the cost of the public. Consequently, economic efficiency cannot be achieved in equilibrium due to 3Ps' wrong adoption as a governance structure for the particular project.

One may argue that governments and developers will gradually learn from experiences and eventually reach the desired equilibrium. However, governments often show a prolonged learning curve in 3Ps. Because of the slow learning curve, 3Ps' transaction costs will be much higher due to the off-equilibrium impacts. We argue that three primary reasons are contributing to the slow learning curve. First, governments suffer from bureaucracy's rigidness and have limited incentives and flexibility in adjusting their practice or standard procedure. Sometimes, even when a particular bidder may be the best one for the project, governments cannot favor this bidder but can only follow the standard procedure and use standard bid evaluation criteria and weights for procurement. Second, each 3Ps project is unique, considering the different possible situations, such as project types, locations, local governments involved, financing agencies, bidders, political environment, economy, etc. Therefore, it is challenging for governments or even scholars to accurately and objectively determine the significant causes of project failures and learn lessons. As a result, governments often learn little from past experiences in 3Ps. Third, 3Ps are often misused as a tool by governments for boosting the economy for governments' re-election purposes. They may be too soft on developers in *ex-ante* negotiation or *ex-post* renegotiation to provide more infrastructure in a short period. Even worse, if a project succeeds, the government takes the credit and, if the project fails, the consequences will likely be borne by the new government in the next term. Therefore, as we have observed, 3Ps have often been applied to those projects that can be easily procured by traditional delivery methods or those that are not in great need. Thus, governments seldom admit these mistakes they have made and will not be able to learn. While governments have a slow learning curve, developers have a quick learning curve due to market competition and their profit maximization objective. When the two distinctive learning speeds combine, it will be optimal for developers to engage in opportunism instead of sending signals that will not be correctly taken.

2.5 A contingency theory of 3Ps as a governance structure

As shown in Figure 1.8, the interaction dynamics between the public and private sectors affect each party's evaluation of the outcomes of a Public–Private Partnership project, namely, social gains, transaction costs, and the value of profit pool. More specifically, from the game-theoretic perspective, whether 3Ps are a good governance structure for a particular project depends on many factors, primarily centering on the potential transaction costs due to unbalanced profit structures and renegotiation/hold-up problems. These factors can be categorized into project factors and institutional factors, as summarized in Table 2.1. A more detailed discussion is given as follows.

Project factors that may cause high transaction costs due to an unbalanced advantage structure may include the following:

- The profit structure tends to skew large construction contract returns severely. For example, when the project scale is enormous, the equity ratio is meager, and the primary Developer is the future contractor.
- The profession of the primary Developer is not connected to future operations. This is usually a sign of a potential unbalanced profit structure. On the other hand, when the major Developer's profession is directly related to the future operation, such as the most potent plant 3Ps, there will be much less concern for an unbalanced advantage structure.

Table 2.1 Negative elements for 3Ps as a governance structure

Project factors	Institutional factors
• Project scale: too large to fail [R] • Project importance: too essential to fail [R] • Project complexity: too difficult to replace the incumbent firm [R] • Profit structure: skewed to large construction contract returns [U] • The profession of the Developer: not related to future operation [U] • Uncertainty of future tolls/revenues: too high to have a reasonable forecast [U]	• Government's professional capability in 3Ps: inexperienced [U, R] • Fiscal market: immature [R] • Government's tolerance for project failure: low [R] • Legal system: immature [R] • The legitimacy of government subsidies: high [U, R]

Notes: [U]: factors contributing to unbalanced benefit structure; [R]: factors contributing to renegotiation/hold-up.

- The uncertainty of future tolls/revenues is too high to have a reasonable forecast. For example, the revenue forecast on rail/high-speed rail projects or those unfamiliar, first-time projects is usually very imprecise. For this reason, developers will heavily discount their long-term equity and operation returns and then inevitably focus on other short-term returns. The institutional factors associated with an unbalanced advantage structure may include:
- The inexperienced government does not have the professional capability in 3Ps. In this situation, the government may not realize the problem and associated costs of opportunism due to an unbalanced profit structure.
- The legitimacy of the government's subsidy for capital needs is high. In this scenario, the equity investment may be much lower than usual and, thus, will cause an unbalanced profit structure.

Project factors that may cause high transaction costs due to renegotiation/hold-up may include the following:

- The scale of the project is too large to fail. In this case, the government will be more likely to be held up for rescue if the project fails. In practice, few governments will allow mega-3Ps projects to fail.
- The project is too important to fail. When the project is politically essential or has a particular symbolic meaning to the government's performance, the project is usually not allowed to fail.
- The project is too complicated such that it is too costly to replace the incumbent Developer. For projects with very detailed know-how in construction or operation it is often challenging to replace the incumbent Developer; thus, the government can be easily held up for renegotiation.

The institutional factors associated with an unbalanced advantage structure may include:

- The inexperienced government without professional capability in 3Ps can also cause a hold-up for renegotiation. For example, ignorant governments may tend to use 3Ps for large projects or essential projects.
- When the fiscal market is immature, there will be few alternatives for handling a distressed project except the government's subsidy. In developed countries with mature fiscal markets, project distress problems are often resolved through market mechanisms, e.g., the two rounds of fiscal restructuring in the Channel Tunnel.
- When the government has a low tolerance for project failure, the Developer may sense the government's attitude and hold-up opportunity.
- Like an immature fiscal market, an immature legal system can provide few alternatives other than government subsidies to resolve project distress.
- Lastly, government subsidies' legitimacy will form the Developer's *ex-ante* expectation on the opportunity for renegotiation/hold-up.

2.6 Remarks

From the perspective of the 3Ps profit structure, the original developers/share-holders controlled the costs of the construction at $14.1 billion. In comparison, only $0.89 billion was injected via the original investors into the equity of Taiwan high speed rail. Therefore, given such an unbalanced advantage structure, the controlling investors/developers naturally had the incentive to overestimate the traffic demands to win the concession, recoup their investment, and let the passive stakeholders suffer the losses in the operation period. The developers were seriously criticized that they had recovered most of their investments from undertaking the project's construction contracts.

The Taiwan high speed rail project demonstrated how transaction costs were incurred due to unbalanced profit structures. When the developers identified rent-seeking opportunities due to the project's unstable profit structure, they would bid very aggressively using an overly optimistic fiscal projection to win the project. The actual traffic demands of Taiwan high speed rail during the operation were only half of the original estimates in their bids, and the projects were both delayed by a year or so. Since the project was awarded to the businesses via the most unrealistic forecasts, they soon faced fiscal difficulties either during the construction or early operation stages. The Taiwan high speed rail has made excess profits during government rescues and undertaking the project's major contracts. The failure of the project and the ineffective project execution created substantial transaction costs for the government.

Moreover, we found that this project exhibited project and institutional characteristics discussed earlier in Section 2.3, leading to an unbalanced 3Ps profit structure. The THSR project was unique and the first of its kind in the country. Consequently, the information asymmetry between the government/public and contractors was notable and generated excess construction advantages and the contractors' incentives to become the controlling developers. Also, since the project scale was so big that investments from passive shareholders were needed, the contractor-developers could limit their equity investments to relatively smaller proportions. As a consequence, the Taiwan high speed rail project was prone to a highly unbalanced profit structure, leading to subsequent project failures and transaction costs. From the governance structure perspective, this case study showed many substantial disadvantages in adopting 3Ps as the governance structure for projects exhibited the characteristics in Table 2.1.

From the renegotiation/hold-up perspective, taking over the government's project involved substantial costs because of its original debt guarantee for the first syndicate loan. If the concession contract was terminated early, the government must assume the outstanding debt of the first syndicate loan, around $8.45 billion, according to Taiwan high speed rail annual report of 2008. As a result, the government was subject to hold-up and intended to grant subsidies to Taiwan High Speed Rail in the expectation of turning the situation around and delaying the project's insolvency.

We also found that this project exhibited some characteristics that might cause governments to be held up. For instance, since this project was one of the largest Public–Private Partnership projects globally, the political and economic impacts of project failures were too high, and the government helped up for renegotiation. In contrast, in the more matured institutional environments such as the UK and France, the inefficiency caused by held up was well recognized, and the public had minimal tolerance for renegotiation. For instance, in the Channel Tunnel project, the two governments made it explicit that no subsidies could be granted in any case. Again, this case showed many substantial disadvantages in adopting 3Ps as the governance structure for projects exhibited characteristics shown in Table 2.1.

In Public–Private Partnerships, conflicts and strategic interactions among 3Ps developers and governments are widespread and play an important role in 3Ps. Many complicated issues, such as opportunism, negotiations, competitive bidding, and partnerships, challenge both governments' and developers' wisdom. Thus, game theory, concentrating on the strategic interactions and economic behaviors, is very appealing as an analytical framework to research the business and dynamics among the Public–Private Partnerships and form proper strategies for both governments and developers. In this chapter, the game theory modeling method is utilized for analysis. Mainly, we focus on the opportunism problems and the determinants of Public–Private Partnerships as a governance structure. Through game theory modeling, these problems are abstracted to a level that can be analyzed. Moreover, new insights or theories for the regarded issues are developed during the solving of game models. These new theories can assist practitioners, including governments, developers, bankers, etc., to better coordinate via greater efficiency and effectiveness.

This chapter first identifies two major opportunism problems commonly seen in 3Ps: the unbalanced profit structure problem and the renegotiation/hold-up problem. These two problems contribute to the major transaction costs in 3Ps, related to transactional hazards and inefficiency. The magnitude of these transaction costs has critical impacts on whether 3Ps are a suitable governance structure for a specific project. Second, we present two approaches that aim to restore efficiency due to the opportunism in 3Ps. In the first approach, the developers try to send signals to the government to signify that they are long-term profit-oriented developers. In the second approach, the government uses evaluating strategies to discourage opportunistic, short-term profit-oriented developers from participating in 3Ps projects. The effectiveness of some famous or potential signaling and evaluating strategies is also discussed. Third, the contingency view of Public–Private Partnerships as a governance structure is presented. We argue that the slow learning curve of governments and developers' fast learning curve tend to check the equilibrium interactions and, thus, limit signaling and evaluating strategies. Thus, the focus of whether 3Ps can be a good governance structure for a particular project turns to those factors that affect the propensity of opportunisms in an unbalanced advantage structure and renegotiation/hold-up.

We believe that governments can benefit from the proposed contingency framework by avoiding 3Ps' use when the project and institutional factors predict possible significant transaction costs caused by opportunism. By doing so, higher efficiency and a higher project success rate will be achieved. Simultaneously, when 3Ps are used in the right projects, the long-term profit-oriented developers will benefit from the reduced pressure from the opportunistic developers and, thus, have a higher willingness to participate and have better performance.

2.7 Corruption in big 3Ps: there is the elephant in the room!

Notwithstanding the relevance of corruption in project selection, planning, and delivery, the project management context pays little attention to this important phenomenon. This chapter provides the background for the argument regarding selecting, planning, and delivering infrastructure in corrupt project literatures. It presents the different kinds of bribery and the characteristics of projects that are more likely to suffer. Corruption is specifically relevant for large and uncommon projects where the public party acts as a client/owner or even the main contractor (Heydari et al., 2020). Big 3Ps are 'large unique projects,' where public parties play an important role and are very likely to be impacted by corruption. Corruption worsens both cost and time performance and the profits delivered. This chapter leverages the institutional theory to illustrate the concept of 'corrupt project context' and, utilizing the research on Italian high-speed railways, indicate the effect of corrupt literature on big 3Ps.

Corruption is one of the critical problems for public policies. It is one of the major impediments to developing emerging countries and further improving the quality of life in developed nations (Treisman, 2007; Loosemore and Lim, 2015; Locatelli et al., 2017; Tabish and Jha, 2011). The eradication of bribery is one of the critical challenges that the world engages with. Researchers (e.g. Auti and Skitmore, 2008; Akbar and Vujić, 2014) agree that corruption might be eradicated via enhancing education and cultural changes leading to a better government capable of producing policies to tackle this problem. Based on Rose-Ackerman (1996), government policies can reduce corruption by 'increasing the advantages of being honest, enhancing the probability of detection and punishment, and enhancing the penalties levied on those caught ... Such measures generally need substantive law reform and the introduction of more transparency' (p. 47). Tabish and Jha (2012) demonstrate a positive correlation among 'corruption-free indicators' and professional standards, transparency, the fairness of punishment, procedural compliance, and contractual compliance. Vee and Skitmore (2003) extend the view and indicate that ethical behaviors in the construction industry are promoted via ethical guidelines and policies of private organizations and professional bodies and public-sector procurement agencies' leadership. Recently, Kenny (2012) indicated transparency in public procurement as a crucial practice for combating bribery.

Unfortunately, achieving all these fundamental procedures and cultural changes might take decades, while projects need to be continuously planned and delivered. Therefore, while the sociological and political communities cope with the long-run issues, such as the cultural and policy changes, the project management community shall face corruption in projects without further hesitation.

Numerous elements can undermine projects' performance, like complexity or 'technological sublime,' weakness in organizational design and availabilities, optimism bias, strategic misinterpretation, or even specific project characteristics, etc. (Locatelli et al., 2014; Garemo et al., 2015). Corruption shall be one of these elements, but surprisingly it is not regarded in the project administration literature. Based on an inquiry on Scopus in May 2016, only three articles published in influential project management journals[2] (i.e., the *International Journal of Project Management*, the *International Journal of Project Organisation and Management*, and the *Project Management Journal*) have the word 'corruption' in either the paper title, abstract, or keywords. These papers are:

a. Sonuga et al. (2002), which describes bribery, inadequate funding sources, and price variation as major elements that cause projects to fail in Nigeria.
b. Ling et al. (2014), which undertakes a comparative evaluation of drivers and obstacles to accepting relational contracting practices in public construction projects in two different markets: Beijing and Sydney. The authors underline that this type of contract might cause allegations of corruption.
c. Bowen et al. (2015), which analyses the effect of bribery on the South African construction industry.

In project administration, bribery is the 'elephant in the room' that must be acknowledged and argued. This chapter summarizes the critical aspects known from the comprehensive literature concerning such an 'elephant,' demonstrates the relevance of this subject in project management, and offers a research agenda.

As discussed later, corruption is particularly relevant for big 3Ps due to their intrinsic characteristics. Big 3Ps are projects characterized by large investment commitment, vast complexity (especially in organizational terms), and long-run effects on the economy, the environment, and society (Brookes and Locatelli, 2015). Big 3Ps and their literatures are mutually interdependent since they influence each other (Miller and Lessard, 2000). Based on Kenny (2006), 'the major effect of corruption in infrastructure is generally going to be on what is built where not how much is paid to build or connect it' (p.18). Therefore, the investigation of bribery in projects and megaprojects requires considering the mutual interlink via the project literature, which is generally, and prevalently, dominated by public policy and the public procurement framework. The concept of 'corrupt project context,' as illustrated by this section, is useful to this investigation.

The topic of corruption is highly controversial. Moreover, the definition of corrupt project literature is specifically challenging and, to some extent, contradictory; this occurs because, in legal terms, the concept of corruption applies to physical persons as a matter of penal liability. However, pragmatically, it is mandatory to attribute the idea of corruption to complicated socio-economic systems (e.g., organization, country, etc.). Therefore, this section defines a corrupt project literature as an environment where corruption is endemic.

To illustrate the concept of 'corrupt project context,' this research leverages the institutional theory, which provides a flexible and adaptive way of conceptualizing institutions. Based on the perspective of Henisz et al. (2012) and Scott (2005), institutions can be conceptualized by adopting three main perspectives: regulative, normative, and cultural-cognitive. These perspectives permit identifying shared rules, norms, values, opinions, and understandings that characterize institutions.

From one side, institutional theory identifies the project literature (Müller et al., 2015; Scott, 2012; Winch, 2000a). On the other hand, it permits the investigation of corruption as a social/institutional phenomenon instead of an individual crime (Williams et al., 2015; Hauser and Hogenacker, 2014; Shleifer and Vishny, 1993; Uberti, 2016). For pragmatic reasons, this research assumes that bribery is institutionalized at the country level as a social phenomenon; this is justified due to sufficient stability and uniformity of rules, cultural values, and shared perspectives. The country-level consideration is consistent with Jensen and Jr. (2000). It is also valid for the project literature; similar research confirmed its tendency as a reliable institutional context. For example, a description of Germany and Great Britain's project context is provided respectively by Bremer and Kok (2000) and Winch (2000b); Ping (2013); Locatelli et al., (2017); Heydari et al., (2020); Transparency International (2014). With all its scandals and endemic corruption, the Italian case is illustrated in Bologna and Del Nord (2000) and further detailed later in the chapter.

As a sequence, the overlap of these two institutional levels (i.e., project literature and bribery as a social phenomenon) originates from the concept of 'corrupt project context.' This conceptualization implies a significant challenge to evaluating and quantifying the extent to which a socio-economic system is endemically corrupt. This challenge lies in the ability to demonstrate and quantify the actual presence of corruption.

Since often corruption cannot be directly evaluated, this chapter considers (through a case study) two drivers of the understanding of corruption specifically in the project literature, i.e., the indexes of corruption and the historical beliefs. Therefore, the chapter focuses on public corruption, specifically in relation to public policy and public big 3Ps. Doing so paves the way for this research stream in project management by providing relevant and updated background. In particular, the chapter concentrates on two questions:

- *Q1: which project characteristics favor corruption?*

This first question is crucial to understand if attributes make the projects more likely to suffer from corruption. The answer to this question is necessary, specifically for decision-makers (DMs) and policymakers in corrupt countries. For instance, let us assume a 'functional objective' (e.g., provide a certain amount of electricity in a particular area) that can be satisfied via two different projects, types A and B, and one of these (e.g., B) is more likely to attract corruption. Then, based on this criterion, A shall be the right choice in a corrupt project literature.

• *Q2: how does a corrupt context impact on project performance?*

Since projects might have low schedule and budget performance even in 'non-corrupt countries,' this question highlights the effect of bribery behavior by comparing similar projects in different nations.

This part of the chapter methodology is designed to answer the two research questions (RQs) previously presented. Q1 is responded to with a critical literature review (Section 2.8) of sources, mostly outside the project management domain. Q2 is addressed via the review of the related context and further investigated via the Italian research case in Section 2.12. First, the review of the past literature enables us to formalize the key constructs useful for this section: corruption, project context, bribery in project literature, megaproject, etc. Second, it permits us to answer the first question with a list of critical drivers demonstrating the typologies of projects that are more likely to involve corruption. Third, it demonstrates the impact that corruption in projects has on their performance during their lifecycle. In summary, this chapter highlights two main aspects: the drivers of bribery and the impact that corruption has on project performance.

A case study considering a megaproject in a 'highly corrupt context' integrates the context. The case study aims to shed light on the role of the literature of project performance. On the one hand, corruption often specifically exists in big 3Ps; on the other hand, big 3Ps are often related to poor performance, even in countries with low bribery signals. Therefore, the methodology compares the big 3Ps involved in the high-speed rail programs in Europe and globally. The comparison considers two main opinions: first, the project literature and the extent to which it is corrupt, and second, the megaproject's performance, normalized and adjusted to consider different environmental, urbanistic, and technological circumstances. The comparison is *improved around* the research case of the Italian high-speed rail program. This case is utilized as a source because it is delivered in corrupt project context and is technologically comparable to the other European high-speed rail programs. The research case is designed to highlight the connection between the endemic phenomenon of corruption and lower project performance; this technique is implemented based on the principles denoted by Brookes et al. (2015) and Yin (2013). The case study comprises three main opinions: the project literature, the longitudinal view over the project lifecycle, and the transversal view based on project performance. The case study is designed in such a way for pragmatic reasons because (1) it is difficult to show the current situation of the corruption in projects and (2) the focus of the direction in the study is not on single corruption episodes, but the project delivered in a 'corrupt project context.' Therefore, the chapter indirectly shows corruption by referring to the project context and displaying the results of

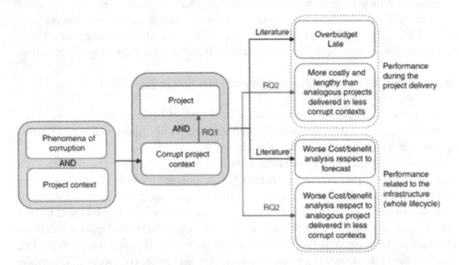

Figure 2.2 Presence of corruption in 3Ps.

Table 2.2 Big 3Ps analysis

Category	Parameter	Threshold
Cost	Cost overruns	>25%
Time	Slip in execution schedules	>25%
Quality	Production versus plan	Reduced production into year 2
Cost	Cost competitiveness	>25%
Time	Schedule competitiveness	>50%

Note: the data about schedules competitiveness are not evaluated because the different length of sectors makes the examination too uneven .
Source: (Merrow, 2011).

judicial processes and the investigations associated with the project. Figure 2.2 summarizes the critical study constructs and their casual interlinks along with the two questions.

In particular, by accepting the research framework from Merrow (2011), this chapter demonstrates how the project and project management performance evolves over the project lifecycle. Merrow's framework evaluates megaproject performance through five parameters. Each parameter is related to the threshold value, which permits us to judge whether the account is satisfactory or not (Table 2.2).

2.8 Corruption the most visible features

Transparency International denotes corruption as 'the abuse of entrusted power for private gain' (Transparency International, 2015a). Based on Aidt (2003), there are three scenarios favoring corruption:

1) Discretionary power: public officials must have the ability to design or administer regulations and policies in a discretionary manner (Ling and Tran, 2012).
2) Economic rents: the manipulation of decisions shall derive some return for the DMs.
3) Weak institutions: the structure of government institutions and the political procedures are fundamental determinants of corruption (Shleifer and Vishny, 1993). Corruption is generally divided into two categories (Transparency International, 2015a): 'petty corruption' refers to everyday abuse of entrusted power via low- and mid-level public officials in their interactions with ordinary citizens; 'grand corruption' refers to acts of corruption committed via relevant institutions like governments and courts. A sub-category called 'political corruption' refers to manipulating policies, institutions, and procedural rules in allocating finances or other resources perpetrated by policymakers.

The Anti-Corruption Resource Centre (Anti-Corruption Resource Centre, 2015) classifies corruption based on the frequency of the phenomenon: 'sporadic corruption' is connected to occasional opportunity; 'systemic corruption' is a mixed and essential aspect of the economic, social, and political systems. The Global Infrastructure Anti-Corruption Centre (GIACC, 2008) demonstrates 47 available acts of corruption during the realization of an infrastructure. These acts are divided into three steps: pre-qualification and tender, project execution, and dispute resolution. Corruption may occur in several ways; the most common are (GIACC, 2014; Anti-Corruption Resource Centre, 2015):

- Bribery, which is committed when a person offers/gives either some advantages to another person, or incentives to act dishonestly. A bribe is not necessarily a cash transaction. It can involve various non-cash benefits for the rogue, such as free holidays, low tenancy fees in prestigious accommodations, etc.
- Extortion: the crime of obtaining money or some other valuables via the abuse of office or authority.
- Fraud: involves the rogue deceiving innocent parties to gain some fiscal or non-fiscal advantage.
- Abuse of power: happens when a person in public office deliberately acts in a way that is contrary to his/her duty and is in breach of his position of public trust.
- Embezzlement refers to the misappropriation of property or funds legally entrusted to someone in their formal position as agents or guardians.
- Conflict of interests: occurs when an individual, via formal accountability to serve the public, participates in an activity that jeopardizes their professional judgment, objectivity, and independence.
- Nepotism: happens when the decision-makers grant favors to their relatives.

These forms of corruption are somewhat similar worldwide. For example, in the South African construction industry, ethical problems include collusion, bribery, negligence, fraud, dishonesty, and unfair practices (Bowen et al., 2007).

2.9 Economic growth and effect of corruption

Sometimes, it is possible to hear unofficial conversations of practitioners and policymakers about the theory of 'efficient corruption.' Historically, even researchers have been debating about the existence of this peculiar concept. Based on this theory's supporters, corruption may play a role as 'grease on the wheel' of economic development, especially where public institutions are weak. Leff (1964) was one of the first writers to support the efficient corruption theory. He tries to overcome the criticism according to moral grounds, and he defines corruption as 'an extra-legal institution utilized via individuals or groups to gain influence over the action of the bureaucracy' (p. 389). Leff (1964) demonstrates that corruption may have a positive effect on economic development because:

- It can make the bureaucrats work harder because grafts motivate them.
- It could avoid the 'red tape,' i.e., the excessive bureaucracy or adherence to rules and formalities, specifically in public business. Hence, the bribe acts as 'speed money,' or rather, it speeds up the bureaucracy.
- It can act as a 'helping hand' to attract FDI.
- It can introduce a factor of competition in situations of closed markets (e.g., natural monopolies). It presents a sort of competitive bidding between the entrepreneurs where those who can perform the work more efficiently are also willing to pay the highest bribe.

Subsequent studies have tried to prove the existence of 'efficient corruption.' Lui (1985) proposes a model where customers can decide to pay bribes for better positions in a bureaucratic queue, for instance, to obtain a license. This model assumes that a person who gives the most outstanding value to time is willing to pay the biggest bribe. The author identifies a Nash equilibrium (NE) that minimize the average value of the cost of time in the queue and maximizes the revenue (in the form of bribes) for the DMs. Beck and Maher (1986) compare an equilibrium model of the graft to a competitive bidding model. They argue that the same firm will win the contract in both cases. Where public policies are weak, corruption could be efficient (Lien, 1988). Egger and Winner (2005) investigate the role of crime as a 'helping hand' for foreign direct investment. This work recognizes as negative impacts of corruption: the payment of bribes, resource wasting in rent-seeking activities, and additional contract risk, while as positive impacts of corruption, it recognizes the following: speed up of bureaucratic processes, and the possibility to access the publicly funded project.

Other researchers have raised sharp criticism against these studies because they do not consider fundamental dimensions in judging corruption's impacts. These researches contend that corruption requires secrecy that makes it distortionary, unmanageable, and costly (Shleifer and Vishny, 1993). Corrupt public officials might deliberately amplify the delay of the bureaucratic system to extort more bribes. The consequence is that nations where corruption and bribery prosper are also those in which the country's firms waste more (not less!)

time with government officials haggling over regulations (Kaufmann and Wei, 1999). Lastly, history demonstrates that when the corruption system is centralized and well organized, the effects on the country's economy are incredibly harmful (Wedeman, 1997).

Mauro (1995) assesses a dataset consisting of indices of corruption, red tape, and efficiency of the judicial system. He finds that corruption is negatively associated with the investment rate, which is connected to economic growth. Mauro (1998) underlines that corruption causes more spending on those components of public expenditure on which it may be more comfortable or much more lucrative to levy bribes.

Consequently, corruption reduces spending on education, one of the major drivers of economic growth. This is a vicious circle since the low education expenditure level might be a reason for further corruption (Mauro, 1998). Mo (2001) asserts that a 1% increase in the corruption level decreases the growth rate by about 0.72%. Since the money paid for corruption might decrease the profit, the construction company might be keen to recover a portion of the advantage via subcontracting at the lowest cost, losing quality and value for money (May et al., 2001). Tanzi and Davoodi (1998) concentrate on qualitative impacts and demonstrate how corruption:

- Decreases public revenue and enhances spending, contributing to the fiscal deficit.
- Increases income inequality because it allows well-positioned individuals to take advantage of government activities at the rest of the population's cost.
- Imposes regulatory controls and inspections for market failures.
- Distorts incentives.
- Performs as an arbitrary tax.
- Decreases the role of government in the enforcement of contracts and the protection of property rights.
- Decreases the legitimacy of the market economy and democracy.
- Acts as an obstacle to entry in the market for small and emerging firms.

Lambsdorff (2003) analyses the connection between corruption and productivity, demonstrating that an increase of corruption by 1 point on a scale of 0 (very corrupt) to 10 (very clean) lowers productivity by 2%. Wei (2000) researches the impacts of corruption on foreign direct investment (FDI), demonstrating that corruption negatively impacts the ability to attract FDI. Tanzi and Davoodi (1998) research the causal connection between corruption in projects and the quality and cost of infrastructure. Based on them, corruption is associated with (1) higher public investment, (2) lower government revenue, (3) lower expenditures on operation maintenance, (4) lower quality of public infrastructure. Furthermore, corruption decreases growth by (1) higher public investment while decreasing its productivity, (2) lower quality of existent infrastructure, (3) lower government revenues needed to finance productive spending.

2.10 Corruption in 3Ps

The Bribe Payment Index via Transparency International reports the likelihood of bribery in 19 different business sectors. Transparency International (2011) showed that corruption is perceived to be standard across all industries: no section scores above 7.1 (on a 10-point scale, where 10 means meager chances of paying), but the worst one, by far, is 'public works contracts and construction' with a score of 5.3. Furthermore, firms operating in public works and the construction sector not paying bribes are more likely to lose business and see corruption as a deterrent for projects otherwise attractive (Graf Lambsdorff et al., 2004). Not all projects are the same; special features characterize projects and make them more or less susceptible to corruption (Stansbury, 2005). These characteristics are:

- **Size**: this is the most important feature because it is easier to hide bribes and inflated claims in large projects than in small projects.
- **Uniqueness**: this makes budget costs challenging to compare, and therefore they are easier to inflate.
- **Government involvement**: public administrators can use their arbitrary power, specifically where there are insufficient controls on how government officials behave.
- **Some contractual links**: each contractual link provides an opportunity for someone to pay a bribe in exchange for the contract award.
- **Project complexity**: when projects are very complicated, elements like mismanagement or poor design can hide bribes and inflate claims.
- **Lack of frequency of projects**: winning these projects may be necessary to contractors' survival or profitability, providing contractors with an incentive to use bribes.
- **Work is concealed**: subsequent procedures cover the essential components of the work. The quality of the ingredients can be very costly or difficult to check.
- **Culture of secrecy**: even if public funds subsidize the projects, costs could be kept secret.
- **Entrenched national interests**: the government selects local and national companies justifying the choice to favor national interests. These positions have often been cemented with corruption.
- **Lack of 'due diligence'**: frequent lack of due diligence on construction project participants allows corruption to continue.
- **In several cultures' bribery and deceptive practices, the cost of integrity is** often approved as the norm: not paying these bribes means not doing the project.

Zarkada-Fraser and Skitmore (2000) show that stakeholders more susceptible to bribery are younger, not affiliated to professional bodies, and less loyal to their organization, with specifically lower levels of job satisfaction. Tang et al. (2012) analyze the role of corruption in winning contracts on the international market. In particular, procurement is critical; hence corruption can be related to (Søreide, 2002):

- **Invitation**: the public officials may have the power to decide which enterprises to invite to the tender.
- **Shortlisting/pre-qualification**: limiting the number of competitors according to previous experience.
- **Technology choice**: aiming to needs special characteristics for the tender.
- **Confidentiality of information**: there are several ways for public officials to hold confidential information to misuse their position.
- **Deviation from the public competitive procurement process**: for instance, in case of emergency.

The 'megaproject' is a special class of projects that share most of the characteristics above. Big 3Ps are large investment projects that tend to be massive, indivisible, and long-run artifacts, with investments taking place in waves. Megaproject 'effects are felt over many years, specifically as auxiliary and complementary additions are made' (Miller and Lessard, 2000). Public policy robustly impacts the performance of public big 3Ps. Big 3Ps 'remain under political scrutiny well after the final official decision is made. Decisions made early on can have disastrous impacts when abstract political ambitions crystalise in unique technical challenges' (Giezen, 2012, p. 782). Despite their fundamental economic and social role, big 3Ps are often implemented after a weak (or, at least, not optimal) phase of project planning, leading such big 3Ps to failure. Indeed, Flyvbjerg et al. (2003) indicate that public infrastructure big 3Ps are impacted by cost overrun and delays in different phases of the project development, and their operating results do not justify the implementation of the project.

Tanzi and Davoodi (1998) are among the first to investigate the link between corruption and the low quality of the results delivered. Gillanders (2014) indicates that regions with higher crime than the national average tend to have worse infrastructure than others. Van de Graaf and Sovacool (2014) show that corruption can be a source of project failure, specifically in highly corrupt nations. Ma and Xu (2009) describe two major acts of corruption:

(1) To gain unlawful the qualifications during bidding and tender.
(2) To raise prices or reduce the quality of engineering standards during construction.

Therefore, corruption causes at least two major impacts, market distortion, and worse cost/benefit. Considering market distortion, decision-makers may prefer to situate projects in locations under the physical control of particular corrupt officials to enforce corruption (Kenny et al., 2011). Furthermore, decision-makers may push for the use of complicated technologies that need non-standard procurement (see the case of TAV in Section 2.11.2). Flyvbjerg and Molloy (2011) show how costs, time, and advantages forecasts are deliberately and systematically over-optimistic to promote a project at the expense of another. In exchange, politicians may gain either bribes, support for their election campaigns, or both. Therefore, corruption impacts on project and big 3Ps performance leading to the delivery of works with limited social advantage, low economic returns, and budget overruns (Wells, 2014), and building poor-quality infrastructure in the wrong place. Corruption affects the project's quality, starting from the project preparation, and it continues during its implementation via significant acts of corruption (Wells, 2014).

In overall, corruption negatively affects project performance because:

- It delays delivery times and increases infrastructure costs.
- It decreases the potential economy of infrastructure because sub-optimal projects are implemented.
- It decreases efficiency, favoring construction firms with corrupt connections rather than the most efficient ones.
- It reduces the quality of infrastructure services.
- It enhances the operating cost of providing a given level of infrastructure services. It limits access, specifically for the poor, because of the higher price of service related to the higher construction, operation, and maintenance costs.
- It favors the generation of monopolies and market concentrations.

2.11 Literature and choosing a 3Ps

This section provides the rationale for selecting Italy as the project context and (Treno adAlta velocità," i.e., "High-speed train") as a representative project. Based on the statistics presented in Section 2.12.1, Italy seems one of the most corrupt EU nations with crucial concerns in the 'anti-corruption agency' and 'public sector.' Numerous infrastructure projects have been delivered in Italy, and most of them were represented as impacted by corruption. The availability of public data about judicial proceedings and project characteristics in Italy allows the role played by corruption in big 3Ps to be studied. Among other big 3Ps delivered in Italy, this section concentrates on Treno Alta Velocità SpA. However, Treno Alta Velocità SpA is the most extensive infrastructure investment program in Italy over the past 20 years, and evidence of corruption has emerged. Therefore, consistent with the framework in Section 2.7, the current chapter of planning and delivery Treno Alta Velocità SpA scenario in Italy is an exemplary scenario of megaproject(s) delivered in a corrupt project context.

2.11.1 European CPI

The Corruption Perceptions Index (CPI) measures the perceived levels of public party corruption in 175 countries (Transparency International, 2014). In the last report (2014), Italy was 69th worldwide with a score of 43/100 (where 100 indicates a shallow perception of corruption). It was the worst in the EU (EU-15 and EU-28).[3] Figure 2.3 indicates that in the last ten years, the Italian level of CPI has not undergone improvement. Recently Italy became the most corrupt nation in the European Union. The Control of Corruption Index (Figure 2.4) measures the extent to which public power is exercised for private gain and the strength and effectiveness of a country's policy and institutional framework to prevent and combat corruption (World Bank, 2013). In the last survey (2014), Italy occupied the 94th position out of 209 nations (one of the worst in the EU). The lack of practical national anti-corruption efforts in Italy is also underlined by another index improved by Transparency International, the National

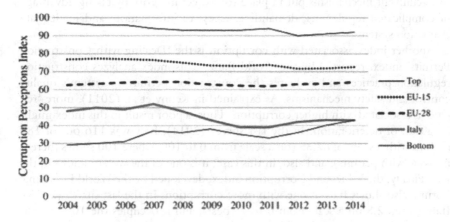

Figure 2.3 Corruption Perception Index trend in the EU and Italy. Elaboration from Transparency International, 2014.

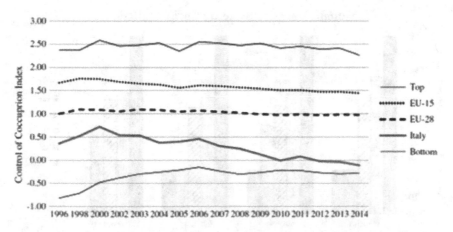

Figure 2.4 Control of Corruption Index in the EU and Italy. The 2014 Italian score was −0.11 on a scale from −2.5 to 2.5, where higher values correspond to less corruption. Elaboration from World Bank, 2013.

Integrity System. The index analyzes the critical pillars in a country's governance system, both considering their internal corruption risks and their contribution to fighting corruption in society, on a scale from 0 to 100. Italy's average National Integrity System score is 57.25, a low value (Transparency International, 2015b). Based on this index, the three weakest pillars in Italy are 'media' (score 38), 'anti-corruption agency' (38), and 'public sector' (42). In contrast, the strongest pillars are 'supreme audit institution' (79), 'judiciary' (75), and 'electoral management body' (75). Comparing the principal pillars (politics, society, culture, and economics) with EU-15 (Figure 2.5) demonstrates that Italy is far from European excellence and well beneath the average of EU-15, specifically in politics and culture. Transparency International (2013) affirms that in Italy, it is easy to circumvent mechanisms put in place to protect integrity by taking advantage of complicated regulations, demanding access to information, and inadequate evaluation systems.

Another index associated with corruption is the 'Dealing with Construction Permits' index, reported by the World Bank. This index analyzes construction regulation practices and evaluates the construction permitting system's quality control and safety mechanisms. As explained in Kenny et al. (2011), more 'red tape' is associated with higher corruption. Thus, a poor result in this index might be a sign of deterioration. In the last survey (2014), Italy was 116 out of 189 nations with a score of 67.35 (on a scale from 0 to 100, where 100 shows there is no issue with permits), and also, in this case, it is one of the worst nations in the EU. Finally, the Ethics and Corruption index, developed by the World Economic Forum, also shows Italy as one with more corruption. In the last survey (2015), Italy scored 2.8 on a scale from 0 to 7 (best), and it occupies the 102nd position out of 144 countries worldwide and the second to last position among the European Union nations (World Economic Forum, 2015). All these indicators

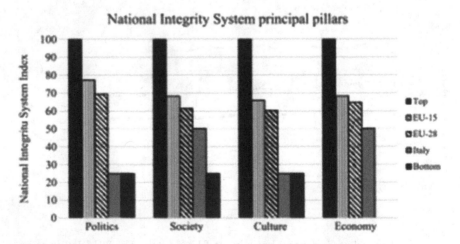

Figure 2.5 National Integrity System principal pillars. Elaboration from Transparency International, 2015b. Data from the European Union not available.

highlight Italy as one of the most corrupt nations in the EU and one of the most corrupt among the developed countries. Sadly, Italy is therefore ideal for studying the problem of the 'corrupt project context.'

2.11.2 Evidence of corruption in Treno Alta Velocità SpA

This subsection highlights the signals of corruption during the big 3Psbig 3Ps' planning to the TVA project. Doing so shows the major investigations associated with crime in the planning and delivery of Treno Alta Velocità SpA. Keeping in mind the idea of 'corrupt project context,' there are four essential premises:

(1) The purpose of the following evaluation is not to reconstruct in detail the trials for corruption involved in the Treno Alta Velocità SpA big 3Ps but to highlight that several prosecution offices throughout Italy investigated the corruption in the Treno Alta Velocità SpA big 3Ps.
(2) Due to its nature, corruption is secret, and stakeholders generally do not benefit from denouncing acts of corruption, so crime is complicated to detect even for investigators.
(3) All investigations for 'petty corruption' are omitted because of the difficulty of finding official information, specifically concerning the big 3Ps' initial phase.
(4) Formally, the Treno Alta Velocità SpA was private up to 2006, and in Italy, the crime of corruption among private entities did not exist.

Figure 2.6 contains a timeline with a selection of significant events linked to TAV corruption divided into three classifications: inquiries, arrests, and judicial acts. Table 2.3 summarizes the major questions for crime in big 3Ps and their outcome.

2.11.3 Corrupt project context: An evidence from European Countries

The concept of 'corrupt project context' is derived from institutional theory, which considers the regulative, normative, and cultural-cognitive perspectives (Henisz et al., 2012; Scott, 2005). This section concentrates on these three perspectives at the country level to recognize a scenario of corruption and therefore determine the extent to which the project context is corrupt. This section explains such evidence in Italy, which is the project context for Italian case.

Exceptional cases of corruption have characterized the history of Italy since its unification (1861). The first scandals, the 'Manifattura Tabacchi' (1868, in English 'Tobacco Manufacturing') and the 'Banca Romana' scandal (1893, in English 'Roman Bank') (Pezzella, 2011), happened during the monarchical period (1861–1946). During World War I and the following years, the reconstruction of the infrastructure network demanded a substantial new fiscal commitment, and public works reached 50–60% of total state investments. During

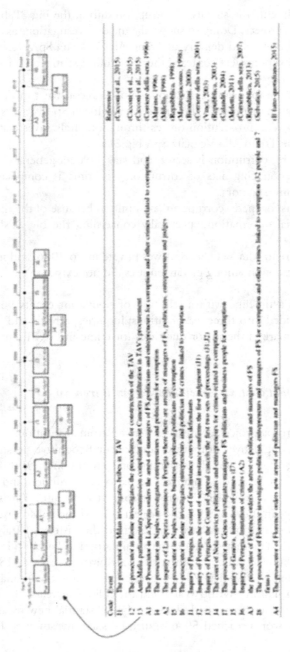

Figure 2.6 Italian case and corruption. Legend Codes: I = inquiry, A = arrest, J = other jurisdictional act. Elaboration from Biondani, 2000; Corriere della sera, 2001; Repubblica, 2013; Calandri, 2004; Corriere della sera, Il fatto quotidiano, 2015, 1996; Corriere della sera, 2007; Marino, 1996; Mastrogiacomo, 1998; Meletti, 2011; Milella, 1998; Repubblica, 1998; Repubblica, 2003; Selvatici, 2015; Vinci, 2003.

Table 2.3 Corruption in construction process and their result

Prosecution office	Defendants	Outcome of inquiry
Rome	Unknown	No further actions
Milan	Unknown	No further actions
La Spezia/Perugia	Politicians, manager FS, and entrepreneurs	Limitation of crimes
Naples/Nola	Politicians and entrepreneurs	Convictions for crime related to corruption
Rome	Politician and entrepreneurs	No further actions
Genova	Politicians, manager FS, and entrepreneurs	Limitation of crimes
Florence	Politicians, manager FS, and entrepreneurs	In progress

the years of fascism, although road works retained a dominant place, the state undertook the realization of a major national program of drying up swamps and the exploitation of waters (Stanghellini, 2000). In this literature, corruption flourished (Bosworth, 2000), and during the 1950s, corruption was widespread, specifically in the local government. The 'INGIC scandal' demonstrated that political parties, even of opposite factions, can collaborate to promote corruption (Camera dei deputati, 1964). In the 1970s, many corruption scandals involved large state-owned enterprises (Almerighi, 1993). During the 1980s, corruption further expanded, becoming systemic and widespread. In the early 1990s, the growing corruption led to a set of inquiries, e.g., 'Tangentopoli' or 'Mani pulite' ('Bride town' or 'Clean hands'), leading to the dissolution of the three major Italian political sections caught with an agreement to manage and distribute bribes (Bologna and Del Nord, 2000). In 1993, the inquiry gained the highest point by discovering the maxi-bribe Enimont/Montedison, getting a value of about €127 million in 2015 equal value. In that case, 4,520 people were investigated, and 1,121 were sentenced. The majority of them quickly returned to public and ordinary life, in part for the statute-barred of the crimes and interest because the political sanctions imposed on politicians involved in corruption scandals were relatively mild (Sargiacomo et al., 2015). Despite the scandal, the Italian legislators failed to prevent the issue of corruption. For instance, the law's reform of public work contracts in the Italian building process following Tangentopoli (Bologna and Del Nord, 2000) had little impact. 'After Tangentopoli, the Italian legislation has not implemented mechanisms to fight against the issue of corruption, but sometimes, has favored its diffusion.' These words come from Raffaele Cantone, of the Italian National Anti-Corruption Authority (Il fatto quotidiano, 2014). Thanks to the proof provided by judicial inquiries, Italy is a model of the failure of common institutional mechanisms to control corruption in an advanced democracy (Vannucci, 2009). In recent years, corruption scandals have become even more common for projects and big 3Ps. Investigators have explored criminal associations sharing public contracts, and the vast majority of big 3Ps performed in Italy (e.g., EXPO, TAV, and MOSE) are under investigation. As also reflected by the indicators later presented, after the period of Tangentopoli, the corruption remained rooted in the system and much more challenging to detect. For example, infrequently, the scandals in recent years are characterized by 'traditional cash bribes.' Conversely, bribes are paid with, for instance, false consultancies jobs.

Table 2.4 summarizes the most relevant Italian projects. The average cost increase is 179% for the whole sample, 216% for railways, and 103% for roads. These numbers are enormous compared to the average increase suggested by the literature.

Table 2.4 Cost overrun for Italian 3Ps from 2011–2014

Type	Type project name	2001 cost (M€)*	2014 cost (M€)	% cost overrun
Rail	Collegamento ferroviario Torino-Lione	1808	4564	152
	Sempione traforo ferroviario	1808	3005	66
	Asse ferroviario Monaco-Verona	2582	9223	257
	Corridoio 5 Lione-Kiev (Torino-Trieste)	7902	30,280	283
	Accessibilità ferroviaria Malpensa	1133	4280	278
	Gronda Ferroviaria Merci Nord Torino	1291	4393	240
	Asse ferroviario Brennero-La Spezia	1511	2766	83
	Asse ferroviario Ventimiglia-Milano	4380	9102	108
	Asse ferroviario Salerno–Catania	12,292	41,149	235
	Asse ferroviario Bologna-Taranto	742	2299	210
	Asse ferroviario Milano-Firenze	1291	13,135	917
	Trasversale ferroviaria Orte-Falconara	1926	3719	93
	Sistema integrato di trasporto nodo Napoli	3886	6624	70
Road	Accessibilità stradale Valtellina	481	2410	401
	Autostrada Cuneo-Nizza	837	3000	258
	Asse stradale pedemontano	3099	9336	201
	Passante di Mestre	2737	4487	64
	Asse autostradale Brennero-La Spezia	1033	4682	353
	Pontina-A12-Appia	1136	4682	353
	Asse autostradale Salerno-Reggio Calabria	13,449	13,843	3
	Asse viario Fano-Grosseto	1854	5119	176
	Asse viario Marche Umbria	1808	2508	39
	Collegamento A1-A14. Termoli S. Vittore	1549	3371	118
	Asse Nord-Sud Tirrenico-Adriatico	1738	4960	185
	Corridoio Jonico	3099	20,171	551
	Nodo stradale e autostradale di Genova	2765	4829	75
Total rail		42,552	134,539	216
Total road		41,225	83,653	103
Total of all projects		78,137	218,192	179

Source: (Prima stima costi all. 1 delibera CIPE 121/2001) Elaboration from (CGIA, 2014).
Notes: * the first cost approximates from the first recondition of the Italian government, 'Interdepartmental Committee for Economic Planning' (Prima stima costi all. 1 delibera CIPE 121/2001). Elaboration from CGIA (2014) and a reference in Figure 2.6.

Also, major events face corruption. Between the last major events held in Italy (the 2006 Winter Olympics in Turin, the G8 meeting in 2009 scheduled in La Maddalena and then moved to L'Aquila, the 2009 World Aquatics Championships in Rome, and EXPO 2015 in Milan), only the Winter Olympics games were not involved in major corruption inquiries. Besides, the structures built for these events (including the Winter Olympics games) are often abandoned at the end of the event, or even worse, never finished, which is a clear symptom of poor planning.

2.11.4 Completed 3Ps in a corrupt solution

2.11.4.1 A case study from Italy

TAV started in 1991 with the announcement of the construction of seven high-speed railways with new technical standards allowing the trains to reach 300 km/h: Rome–Naples (RO–NA), Florence–Bologna (FI–BO), Bologna–Milan (BO–MI), Genoa–Milan (GE–MI), Milan–Turin (MI–TO), Milan–Verona (MI–VR), and Verona–Venice (VR–VE). These new lines were also built to decrease the traffic on traditional lines, allowing better freight and regional trains (Senato Della Repubblica, 2007). In 1998 the project expanded its scope: from an AV system to an AV/AC system (high-speed/high-capacity) to support freight trains (Senato Della Repubblica, 2007). From 1997, these Italian projects were combined with a more extensive European TEN-T program (European Commission, 2015). In 2001 five new lines were proposed: Torino–Lyon (Italian part), Verona–Munich (Italian region), Salerno–Catania, Naples–Bari, and Venice–Trieste (Beria and Grimaldi, 2011).

In 2015, after 24 years, only 4 of the original railways were complete: RO–NA, FI–BO, BO–TO, TO–MI. The subsections Milan–Treviglio and Padua–Venice were built with traditional standards and mixed with HS railways. The railways GE–MI and MI–VR still have sections under construction. The rest of the railways are still in the design phase (RFI, 2015).

The procurement of major infrastructure is always important for corruption; even

> fixed criterion weights ensure objectivity and decrease the risk of unfairness and corruption in evaluating bidders' proposals, but only provided they accurately reflect the relative importance of the analysis elements to the owner. However, it is still available to generate an unfair evaluation system in which too much emphasis is placed on certain evaluation factors, thus favoring (intentionally or unintentionally) those bidders that score highly in the corresponding element.
>
> (Ballesteros-Pérez et al., 2015)

The procurement of the Treno Alta Velocità SpA big 3Ps has been highly controversial and, to some extent, unclear. Cicconi et al. (2015) highlight major dysfunctions in perpetuating the public objectives, specifically in transparency, market freedom (i.e., selecting the contractor, technology, etc.), and fiscal fairness. Fairness is intended, e.g., giving the Italian citizens a fair and adaptable fiscal situation with the public debt and giving the contractors (especially subcontractors) a reliable return and satisfaction (Masrom et al., 2013).

According to Mario Moretti's declaration (former chief executive officer of Trenitalia) during a parliamentary inquiry (Senato Della Repubblica, 2007), Treno Alta Velocità SpA was initially set up as a private project. At the time of incorporating the Treno Alta Velocità SpA (1991), most shareholders were private sector, primarily fiscal institutions (Cicconi et al., 2015). This initial setup permitted them to avoid the public procurement process, including prescriptions concerning transparency, market freedom, etc. The private procurement process allowed them to pre-select the main contractors (of the various infrastructure sections composing the Treno Alta Velocità SpA program) and the infrastructure technology without an open tendering process (Senato Della Repubblica, 2007). Based on the 'freedom of contracts' (Furmston et al., 2012), the contractual framework was published without a public safeguard typical of private bargains.

Afterward, before the construction phase, the fiscal institutions (i.e., the private parties) did not subscribe to the need for more capital to finance the infrastructure's delivery. Since the acquisition of TAV SpA by Ferrovie Dello Stato (FS – "State Railways") in 1998. The infrastructure has been funded by the capital increase and private debt (about 40% of the entire capital cost) guaranteed by the state (Senato Della Repubblica, 2007). Based on Mario Moretti, the Treno Alta Velocità SpA program suffered extended delays, and cost overruns were caused by, among other elements, the choices made during the private procurement procedure (Senato Della Repubblica, 2007). The lack of an open-ended procurement process causes expensive technical solutions to be selected (Senato Della Repubblica, 2007). Specifically, the high-speed network design for both the passenger and the freight trains appears questionable. Freight trains have higher weights, enhancing the whole infrastructure's cost for various reasons, including limitations on slopes and a larger power supply requirement. This unconventional technical selection is an Italian peculiarity. Almost seven years after completing the first high-speed railway in Italy, no freight trains use these lines (Cicconi et al., 2015).

2.11.4.2 3Ps analysis

The project analysis entails comparing cost overruns, time overruns, and benefit in Italian Treno Alta Velocità SpA operations with the same projects undertaken in less corrupt countries.

I. Cost overruns. The evaluation of the costs considers three levels:
1) The first forecast, i.e., the first forecast with the cost declared during the project's launch in 1991.
2) The second forecast, i.e., the cost recognized with the supplementary contract or the contract's signature between the general contractors and FS.

3) The final cost, i.e., the final cost of the railway, or the last estimate available for those still under construction.

Figure 2.7 demonstrates the cost overrun for each railway. Costs are actualized to March 2015. As evident, all the projects are dramatically over budget and well above the 25% threshold (Merrow, 2011).

II. Slip in execution schedules. Figure 2.8 demonstrates the difference between the time scheduled and the real progress of the railways RO–NA, FI–BO, BO–TO, and TO–MI, the only projects delivered. The work begins with a considerable delay, and the execution schedules are longer than expected. The figure also indicates the overrun in execution schedules.

III. Scope and operation. The scope is analyzed by comparing the number of trains traveling daily on high-speed lines with the minimum number of trains that the line can support. The lines were built to manage the passage of a train every five minutes (Ferrovie Dello Stato, 2015), and in specific, the goal is the passage of 120 passenger trains and 40 freight trains per day in each direction (Cicconi et al., 2015). Table 2.5 clearly shows how Treno Alta Velocità SpA railways are dramatically underused.

IV. Cost competitiveness. The comparison between the Italian high-speed project and other EU high-speed railways is difficult due to the differences regarding the technological designs, i.e., the Italian technology is unconventional compared to the other European Union nations. However, the following analysis compares the Italian railways' costs with the Spanish Alta Velocidad Española (AVE) and the French Train à Grande Vitesse (TGV). Table 2.6 lists the railways used for the analysis.

The relative average cost for the Spanish and the French railways is €18.01 million per km. This value should be increased by 25% to obtain the threshold

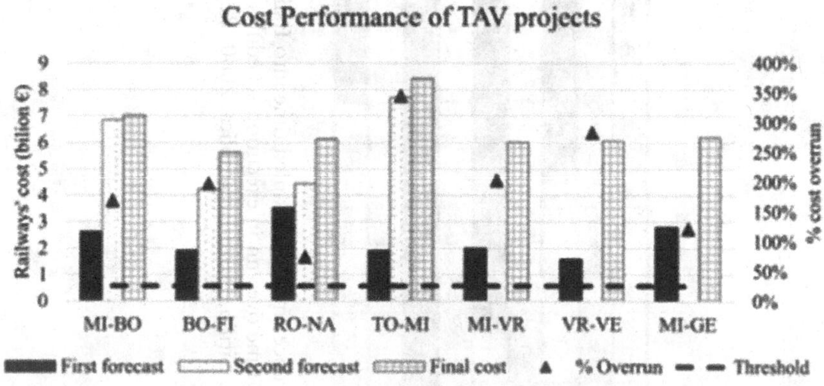

Figure 2.7 Cost overruns for Italian case (currency actualized to 2015). The bar chart (scale on the left – billions of €) indicates the Treno Alta Velocità SpA cost trend (constant currency). The triangles (scale on the right) are the % cost overrun for Treno Alta Velocità SpA railways concerning the 25% threshold. Elaboration from ANAC, 2007; Repubblica, 1994; Camera dei deputati, 2015a, 2015b; Cicconi, 2011; Rossi, 2000.

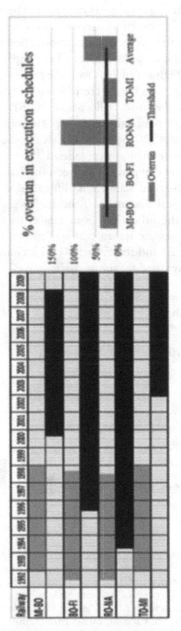

Figure 2.8 Difference among scheduled time into Italian case(grey cells) and real progress (black cells) for completed railways (left). On the right, time overrun in execution schedules plagued TAV railways (right). Elaboration from Repubblica, 1991; RFI, 2015; Oice, 2007; Cirillo, 1994; Repubblica, 2000; Regione Emilia-Romagna, 2015.

Table 2.5 Number of trains for each high-speed railways and percentage of utilisation concerning the capacity of 240 trains per hour

Railways	N° of trains ITALO	N° of trains FS	Total N° of trains	% of utilization
MI–TO	14	28	42	17.5
BO–MI	18	78	96	40
FI–BO	22	126	148	62
RO–NA	16	62	78	32.5

Sources: (Trenitalia, 2015; Cicconi et al., 2015; Italo, 2015)

Table 2.6 Spanish and French railways comparison

Country	High-speed railway	Length (Km)	Opening years	Cost (B€)	Relative cost (M€/km)	Year cost	Cost/km 2015 (M€)
Spain	Madrid–Barcelona	671	2008	8179	12,19	2010	13.07
Spain	Madrid–Valladolid	155	2007	2277	14,69	2010	15.75
Spain	Cordoba–Malaga	201	2007	3729	18,56	2010	19.89
France	Valence– Marseilles	250	2001	4778	19,11	2009	20.76
France	LGV-Est (1)	300	2007	4655	15,52	2009	16.85
France	LGV-Est (2)	107	2016	2010	18,79	2009	20.40
France	Tours–Bordeaux	279	2017	7200	25,81	2010	26.92
France	Le Mans–Rennes	180	2017	3300	18,33	2011	19.12

Costs are actualized to 2015 using conversion rate from http://rivaluta.istat.it/Rivaluta/. Elaboration from Fernández et al., 2012; Ministry of Ecology, 2007; Railway Gazette, 2010; Recarte, 2013; SNFC, 2015.

of failure, so the threshold value is €22.51 million per km. Figure 2.7 compares the Italian railways with the threshold of failure. Also in this scenario, all Italian railways are above the threshold. During an inquiry of the Senate, the management of FS justified the differences in costs between Italian, French, and Spanish high-speed railways with technical elements demonstrated in Figure 2.8. Even accepting the elements and the maximum of the cost increases suggested by FS (i.e., 23 M€/km), the Italian railways' cost remains higher than the Spanish and French railways. Remarkably, as shown in Figure 2.9, Spain and France were less corrupt than Italy throughout the program life cycle. In conclusion, in the parameter 'cost competitiveness,' the megaproject fails, considering planning and territorial differences regarding other projects.

This section covers several case studies of 3Ps projects for the poor in areas like water and telecommunications for various regions (or jurisdictions):

ARGENTINA

Source: 3Ps and the poor – research case – Buenos Aires, Argentina. Experiences with water provision in four low-income barrios in Buenos Aires, Argentina.

A case of four barrios (neighborhoods) in Buenos Aires, Argentina, where the initial water concession from the local government to Aguas de Argentina – the

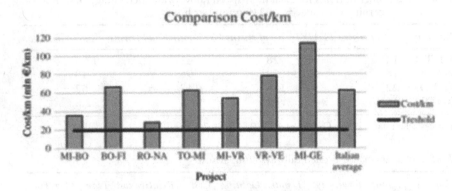

Figure 2.9 Cost/km comparison between European Union HS railways Cost/km comparison among Italian, French, and Spanish HS railways. Elaboration from Railway Gazette, 2010; Railway Gazette, 2011; Fernández et al., 2012; Recarte, 2013; Ministry of Ecology, 2007.

private utility – did not provide enough coverage and left the poor out of the area of service. However, water access and sanitation services improved significantly for citizens, including the urban poor, in an informal settlement. The development resulted from renegotiating contractual terms and the community's strong participation, the government, and the private utility.

Source: universal service obligations in utility concession contracts and the needs of the poor in Argentina's privatizations.

This document is a case study on Argentina's obligatory service and universal service obligation for diverse sectors such as energy, telecommunication, water and sewage, and gas. It opens with an evaluation of the obligatory service and universal service concepts, including the Argentinean experience, and concludes with the case's relevant principles.

BOLIVIA

Source: designing pro-poor water and sewer concessions: early lessons from Bolivia.

In this document, the author describes how the design of laws and regulations, bidding processes, and contracts plays an essential role in private participation in infrastructure. The writer researches concessions for water and sewer services in La Paz and El Alto and describes how the service coverage is directly impacted by the concession framework's provisions. The writer also suggests developing private-sector performance by enforcing well-defined contractual arrangements between the government and private-sector participants.

Source: Expanding water and sanitation services to low-income households – the scenario of the La Paz–El Alto concession.

This section captures the water concession in Bolivia, specifically designed to develop and expand services to the poor. The writer evaluates the effect of changes in contractual provisions where the scope of services is widened.

BRAZIL

Source: case study: São Paulo telecenters project.

This case is about the involvement of Rede de Informacão para o Terceiro Setor (RITS), a Brazilian civil society organization involved in information and communication technology policy monitoring and advocacy – in setting up community access centers (telecenters) in São Paulo, which inspired several policies to roll out telecenters in Brazil. This partnership-associated project mobilized policy, investment, and technical support, leading to 128 community-associated telecenters. The São Paulo model was in accordance with free public access to facilities and training, community participation in management, free and open-source software, and community telecenter development as a social organization venue.

INDIA AND AFRICA

Source: case study: rural broadband backbone (in English, Spanish, and French).

This document discusses different strategies for establishing a broadband network in rural areas. It describes how information and communication technology can help eliminate poverty and why information and communication technology pro-poor policies are essential for the short- and long-run improvement of an economy. It covers several instances of information and communication technology projects in India with direct investment from the government as a public operator and in Africa, where a public–private consortium executed the information and communication technology facility.

INDIA

Source: case study: digital inclusion policies: some lessons from India (in English, Spanish, and French).

This section discusses the Common Service Centres (CSCs) scheme established via the government under the National e-Governance Plan. The plan generates a network of common service centres in India's rural zones, led by the private sector, to promote digital inclusion.

INDONESIA

Source: 3Ps and the poor – research case – Jakarta, Indonesia. Drinking water concessions (research for better understanding 3Ps and water provision in low-income settlements).

This document evaluates several aspects of private-sector participation in the drinking water concessions in Jakarta. The objective of the report is to:

- Study the agreements and resulting outcomes.
- Explain the perceptions of the different stakeholders.
- Explain measures to develop the accessibility of drinking water for the urban poor.
- Discuss options in developing partnerships via community-based organizations.

NAIROBI

Source: 3Ps and the poor – research case – Kibera. Small entrepreneurs and water provision in Kibera, Nairobi.

This is a case study on private-sector participation in the water sector in Nairobi, Kenya, specifically in the city's largest informal settlement, called Kibera. The case analyzes a project developed to re-arrange the existing 3Ps where private-sector participants were small local providers and established water services across informal settings. The research evaluates this particular project's underlying principles and describes why water provision in a Public–Private Partnership is a significant facet of the sector, filling gaps for low-income consumers' effective access to services.

PAKISTAN

Source: 3Ps and the poor – research case – 3: Awami tanks in Orangi Town, Karachi, Pakistan.

This is a case study on water service projects in Karachi, Pakistan, where the Karachi Water and Sewerage Board's performance has declined severely, causing shortages and other problems in the area. Specifically, the report analyzes how, because of a public–private initiative, the water supply situation was improved in Karachi's relatively sizeable informal settlement. The case also draws important conclusions about this specific 3Ps and gives some recommendations about it. The initiative implemented Awami tanks (Awami = communal), facilitated by the Karachi Water and Sewerage Board (KWSB) and Pakistan Rangers.

PERU

Source: research case: providing universal access: FITEL, Peru (in English, Spanish, and French).

This is a critical case on ICT universal access provision; it shows an innovative bidding process that allowed telecom universal access in Peru.

FITEL in Peru offers an early and successful instance of a universal access fund adopting an innovative technique to achieve access in rural areas, now widely replicated: the lowest-subsidy auction. This is an efficient mechanism for minimization. The subsidy needs commercial telecom companies to extend the network into non-commercial areas via awarding the contract to the bidder seeking the lowest subsidy. Despite shortcomings, this pioneering program brought several social advantages, and activities have since expanded from public telephony to include internet access.

Source: telecom subsidies – output-related contracts for rural services in Peru.

This section describes how the lowest-subsidy auction functioned in Peru for the telecommunications sector. The writer analyzes the bidding process, pricing and subsidies, evaluation of the project as a policy, and outcomes.

Source: research case: the Huaral Valley agrarian information system, Peru.

This case is on the Huaral Valley project that, through infrastructure, allowed access to telecommunications and internet service to poor farming communities. It also describes the importance of community empowerment and leadership that significantly improved ICT projects and made them more affordable.

THE PHILIPPINES

Source: the design of the Manila concessions and implications for the poor.

This document studies a Manila program implemented to obtain universal water coverage during its first ten years. The plan was impactful in low-income neighborhoods due to the structure of concessions. The study details the project's background, contractual arrangements, and incentives and concludes with the lessons learned from this case.

SOUTH AFRICA

Source: 3Ps and the poor – research case – Dolphin Coast water concessions, Dolphin Coast, South Africa.

This is a case study on the existing water concession provisions in the Dolphin Coast area in South Africa, including several outcomes before the report was published, different perceptions of several stakeholders, and its impact on low-income groups in that specific area. It is a summary of the evaluation of the 3Ps and its effect.

Source: 3Ps and the poor – research case – revisiting Queenstown, South Africa.

A 3Ps case for the operation, maintenance, and management of existing water and wastewater systems and partnership involve the poor. The 3Ps has been entered into via an operator called '*Water and Sanitation Services South Africa*' and the municipality of Queenstown, of the then-Queenstown Transitional Local Council area (Eastern Cape Province) in South Africa.

IMPROVING NATIONS

Source: contracting out water and sanitation services – Volume 2: research cases and evaluations of service and management contracts in improving nations.

This document starts with an overview of the water sector in Public–Private Partnerships, trends in different sections of the world, and an evaluation of

contracting out and outsourcing. Then the study goes in-depth on cases of contracting out (services and management contracts in the water section) in countries
such as Chile, Mexico, Trinidad and Tobago, Haiti, African, India, Indonesia,
and the USA. The document attempts to answer questions about where and how
to conduct contracting out, whether contracting out has any impact on efficiency
and effectiveness, and how contracting out can be further increased to deliver
developed water and sanitation services in improving nations.

More on Pro-Poor 3Ps at fundamental problems in 3Ps for the Poor, Laws,
and Regulations Supporting Pro-Poor Services Delivery, Contractual instances
Supporting Pro-Poor Services Delivery, Other Mechanisms Supporting Pro-Poor
Services Delivery as well as on Further Reading section.

2.12 Discussion

Table 2.4 demonstrates that the average budget overrun of the 27 Italian infra-
structure projects is globally 179%. Table 2.7 compares the budget overrun of
Italian infrastructure with analogous infrastructure delivered in other countries.
The budget overrun for the Italian infrastructure (including Treno Alta Velocità
SpA) is much higher than analogous infrastructure provided in other countries.

Table 2.7 Summary of the context of budget overrun in the road, rail, and fixed links

Reference	Location	Sector/ infrastructure	Sample size	Overbudget (%)
Table 2.3 of this chapter	Italy	Rail	13	+216
		Road	13	+103
		Total	36	+179
Flyvbjerg et al. (2016)	World	Roads	863	+20
	Hong Kong	Roads	25	+11 + 6/− 1
Cantarelli and Flyvbjerg (2015) updating	World	Rail	58	+34
		Fixed link	33	+20
Flyvbjerg (2008)		Road	167	+34
Cantarelli and Flyvbjerg (2015)	Europe	Rail	23	+34
		Fixed link	15	+43
		Road	143	+22
		Total	181	+26
	North America	Rail	19	+41
		Fixed link	18	+26
		Road	24	+8
		Total	61	+24
	Other geographical areas	Rail	16	+65
Cantarelli et al. (2012)	Netherlands	Road	37	+19
		Rail	26	+11
		Fixed link	15	+22
Lee (2008) cited in Cantarelli et al. (2012)	South Korea	Road	138	+11
		Rail	16	+48

The escalation is even higher than what was observed in billions unique infrastructure such as the Eurotunnel (increased by 59% 'alone' or 69% including the associated projects (Winch, 2013)). Several arguments can explain the budget overrun of infrastructures. Cantarelli et al. (2010) summarize the most crucial ones that emerged from Flyvbjerg's research (see essential reference in Table 2.6):

- **Technical**: forecasting errors including price rises, poor project design, and incompleteness of estimations, scope changes, uncertainty, inappropriate organizational structure, inefficient decision-making process, flawed planning process.
- **Economical deliberate underestimation**: due to lack of incentives and sources, inefficient use of resources, dedicated funding process, poor financing/contract management, strategic behavior.
- **Psychological**: optimism bias among local officials, cognitive bias of people, cautious attitudes towards risk.
- **Political**: deliberate cost underestimation, manipulation of forecasts, private data.

Certainly, optimism bias, strategic misrepresentation (two of Flyvbjerg and Cantarelli's central arguments), and, in case of a program, the escalation of commitment (Winch, 2013) might account for a portion of the budget overrun. However, in Italy and other countries such as France and Spain and arguably everywhere (Table 2.7), the decision-makers were subject to optimism bias, strategic misrepresentation, and other problems. Yet, their budget overrun in projects is in the region of 10–50%. The cost performance of the Italian infrastructure is dramatically worse than elsewhere. Therefore, there shall be something more on top of the list mentioned above of factors (Cantarelli et al., 2010). We discuss that the high level of corruption is likely to play a critical role in selecting, planning, and delivering infrastructure projects.

The budget overrun value of Italian infrastructure is systematically higher, with no relevant improvement over time. This seems an instance of 'tolerance for deviation' (Pinto, 2014), i.e., people within the organization become so accustomed to a deviant behavior that they do not consider it as strange anymore. Normalization of deviance suggests that the unexpected becomes expected. Thus, a deviant culture, such as the tolerance of corruption, becomes approved and causes counterproductive behaviors. More specifically, politicians play important 'political games' and maintain important contacts to ensure broad-based support for the project despite the terrible project management performances (Pinto and Patanakul, 2015). This is a further conceptualization of the temporary project/program organization proposed by Winch (2014), where the permanent organizations are often supported by the Italian state. This implies having the policymakers as salient stakeholders in most critical organizations all over the project and operation phase. A long-term view, embracing the stakeholders over the life cycle, is fundamental for appraising the 'project success' (Turner and Zolin, 2012).

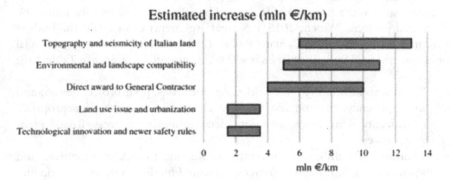

Figure 2.10 Causes and estimating of the higher cost of European Union HS railways. Elaboration from Senato Della Repubblica, 2007.

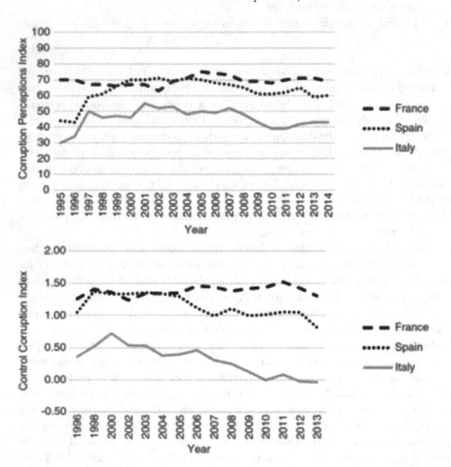

Figure 2.11 Comparison of indexes of corruption between European countries.. Elaboration from Transparency International, 2014, top; World Bank, 2013, bottom.

This chapter discovered the role of corruption in project management (Figures 2.10 and 2.11) and megaproject administration, taking the lead from two ambitious questions in the study. The first question in our study aimed to understand which project characteristics favor corruption. An in-depth evaluation of the current literature allowed us to recognize several features that enhance the odds of a project to suffer corruption, including project size, uniqueness, heavy government involvement, and technical and organizational complexity. We observed that big 3Ps match all these characteristics. Furthermore, projects delivered in 'corrupt nations' or 'corrupt project literatures' are more likely, ceteris paribus, to suffer corruption than in less corrupt literature. Project literature issues: discretionary power of officials, economic rents of policy/decision-makers, and weak institutions make a country ideal for corruption. In this scenario, the investigation of the connection between projects and strategy becomes essential, and the evaluation should be carefully improved from the first step to the end (Pinto and Winch, 2016), which is the 'critical phase' for most of the big 3Ps (Merrow, 2011).

The second question in this chapter investigated how corruption affects project performance. To address this question, the authors performed an in-depth evaluation of the Italian high-speed railway system. The choice allowed a comparison of the Italian scenario with analogous European big 3Ps, providing an explorative but thorough answer to the question of the study. Corruption is harmful to both project administration success and project success. During the project phase, the infrastructure suffers extra costs concerning both its budget cost and other comparable infrastructure.

Similarly, looking at the schedule, there are remarkable delays. During the operation step, the infrastructure fails to deliver the expected advantage, e.g., Italy's high-speed railway system is underused. The Italian state's role as a 'strong owner' (Winch and Leiringer, 2016) is hugely problematic. More generally, corruption leads to a sub-optimal allocation of resources. To carry this low allocation over several projects, 'from a strategic misrepresentation perspective, the lie has to be repeated many times, and as evidence of escalation accumulates, that lie has to be repeated in the face of growing opposition from those who are providing the fiscal resources' (Winch, 2013, p., 726).

Notes

1 A evaluating algorithm model that generates useful information for project choices in a timely and useful fashion at an acceptable cost can serve as a valuable tool in helping an organization make optimal choices among numerous alternatives.

2 Other journals published papers about 'corruption in projects.' This sentence stresses how this topic is under-researched and under-published. Relevant contributions from other journals are quoted in this chapter.

3 The term EU-15 refers to states that joined the EU before 2004, i.e., Austria, Belgium, Denmark, Finland, France, Germany, Greece, Italy, Ireland, Luxembourg, Netherlands, Portugal, Spain, Sweden, and the United Kingdom. EU-28 refers to all current member states of the EU, so it includes the aforementioned EU-15 states plus Bulgaria, Croatia, Cyprus, Czech Republic, Estonia, Hungary, Latvia, Lithuania, Malta, Poland, Romania, Slovakia, and Slovenia.

3 Game theory approach to corruption in Public–Private Partnership[1]

3.1 I invite the minister, who does not? If I don't, who does?

Significant forms of corruption constitute a severe threat to the functioning of societies. The most frequent description of how severe corruption emerges is the slippery slope metaphor – the notion that bribery unfolds gradually. While having widespread theoretical and intuitive appeal, this notion has hardly been tested empirically. We utilized a recently developed paradigm to test whether brutal, corrupt acts unfold gradually or abruptly. The results of three experimental studies revealed a higher likelihood of severe corruption when participants directly faced this option (abrupt) than when they had previously engaged in minor forms of corruption (gradual). Neither the size of the payoffs, which we kept constant, nor evaluations of the actions could account for these differences. Contrary to widely shared beliefs, sometimes the route to corruption resembles a cliff rather than a slippery slope (Xiaohu et al., 2020).

This instance represents a structure of corruption – described in this literature as 'misuse of an organizational characteristic or authority for private or organizational (or sub-unit) gain, where misuse, in turn, refers to departures from approved societal scales' (Anand et al., 2004). Corruption frequently impairs groups, organizations, and communities (Rose-Ackerman, 2006). Experimental investigations into corruption show very damaging social outcomes, for example, a lack of acceptable government development (Mauro, 1995), waste of national wealth (Hellman et al., 2000), and destruction of nature. Extensive studies of corruption in various fields were examined (Bardhan, 1997; Ostrom et al., 1999). In general, more than one significant correlate of corruption is obtained, from lack of transparency (Rose-Ackerman, 1997), to colonial history (Treisman, 2000) to extractive institutions (Acemoglu and Robinson, 2012) that are used for a more clear survey (Rose-Ackerman, 2006; Rothstein, 2011a). Heretofore, corruption research has dedicated an awful lot less interest to psychological elements that assist in explaining why corruption is prevalent in some literature while being almost unavailable in different types of literature (see for some essential exceptions: Darley, 2005; Dungan et al., 2014; Lee-Chai and Bargh, 2001; Mazar and Aggarwal, 2011).

DOI: 10.4324/9781003177258-3

With notice to psychological factors that influence corruption, researchers (Rothstein, 2000) in politics and economics (Lambsdorff and Frank, 2011) have emphasized the importance of realized social factors (Haidt and Kesebir, 2010). Social hierarchies are those beliefs that exist concerning others. As we will argue, corrupt agents can use descriptive indicators to determine the likelihood of corrupt transactions as a decision criterion. Therefore, we attempted to examine the impact of descriptive indicators on engaging in corruption using a single corruption behavioral index in three imperative research types.

In order to realize the way factors affect corrupt behavior, we have to distinguish two fundamental kinds of factors, called descriptive and injunctive factors. Descriptive indicators bring facts associated with how most human beings behave in a specified position. They also express the number of times a particular behavior has been observed. Injunctive scales bring information associated with the specific movements that almost everybody accepts or rejects – hence, whether this particular behavior is suitable and moral (Cialdini et al., 1990; Reno et al., 1993).

In the current section, we focus on how descriptive indicators influence corruption behavior for two fundamental reasons. First, descriptive indicators are subject to inter-community variance. That is, the descriptive indicators associated with corruption differ significantly in a given social literature (Kurer, 2005; Persson et al., 2012; Rothstein and Eek, 2009) – a human has different beliefs related to the abundance of corruption (Lambsdorff et al., 2005). They are injecting scales related to corruption charges into the three social kinds of literature drastically. People often think convergent and corrupt beliefs to be dishonest and false (Karklins, 2005). This moral conviction is also proven in law (Mungiu-Pippidi, 2006; Mungiu-Pippidi, 2017). Corruption is a crime under most national bylaws (Olken and Pande, 2012).

Second, injunctive scales are much less malleable than descriptive indicators. While the views above associated with corruption being incorrect and inappropriate are relatively stable, the beliefs related to the descriptive indicators associated with corruption can be subject to change. Notably, in domains in which humans do not have their personal experience with corruption, the beliefs associated with the abundance of corruption are malleable. Changing descriptive indicators is provided as one of the most promising approaches to fight corruption (Rothstein, 2000).

In general, it can be said that descriptive indicators change in different societies and are generally flexible. In some sociological studies, corruption is seen as pervasive, and other studies rely on the game-theoretical model claim that corruption (Kosfeld, 1997) is almost non-existent (Bicchieri and Rovelli, 1995). Small and large communities can balance corruption limits by considering the results and an abundance of research corruption. The critical point is that the balance must be stable, not usually stable (Bicchieri and Rovelli, 1995). A system can move from excessive descriptive corruption indicators to low descriptive corruption indicators and vice versa (Ashforth and Anand, 2003). To achieve this, corruption-related rules need to be formulated to influence corrupt behavior (Dong et al., 2012; Rothstein, 2000; Bicchieri and Rovelli, 1995).

Imagine bribing a police officer after violating a traffic law. If you believe in venality, you will feel empowered and successful by paying a bribe, such as slipping a note into your driver's license, because you have escaped a heavy fine. In this case, the anticipated value of accepting a bribe by a police officer is higher than the potential penalty. However, if you think this type of corruption is less pervasive and less frequent, this approach may put you at greater risk than previously stated. In this situation, the potential reprimand for trying a payola outweighs the expected value of bribe acceptance. Complementing this example, it has been repeatedly stated that the abundance of corrupt behavior depends on whether one considers others corrupt. This does not mean that one considers it unacceptable or illegal (Dong et al., 2012; Ariely, 2012; Olivier De Sardan, 1999).

In other words, these interpretations suggest that descriptive scales influence corrupt behavior. Previous studies have not yet established this relationship empirically. This was partly due to the lack of appropriate methods to investigate corrupt behavior. For this reason, this study not only researches this problem but also introduces a distinctive corruption game. In this game, we ask participants to imagine that they are the CEO of a development company, and their bridge construction contract is on the desk of an official. Then we ask them to decide whether to bribe the official. Therefore, their behavior will be more influenced by perceived descriptive scales and less by personal experience.

We include a corrupt deal in this economic game frame. In other words, we hide the corrupt motion in the form of a unique, valid public invitation that has commercial profits. On the other hand, they have been explicitly asked to pay bribes or frankly withhold payment. Even members were involved in paying bribes. Masking corruption causes its prolongation, as it reduces its impact on favorable social conditions (i.e., human beings are no longer attracted to corruption because it is socially unacceptable). However, this behavior was found to be "corrupt" because the subjects in all three studies perceived the invitations from the public to be significantly more corrupt than the non-invitation from the public official (all ps < .016)

In this part of the chapter, we set out to evaluate whether descriptive scales – the belief associated with the abundance of corruption in a given context – anticipate corrupt behavior. Using the distinctive corruption game, we performed three studies. First, we examined two correlational studies for whether the realized descriptive scales before (first corruption study) and after (second corruption study) the corruption game correlate with corrupt behavior in the game. Second, to evaluate the causal relationship, we set up an experiment in which facts associated with descriptive scales were manipulated to examine the effect on the subsequent corrupt behavior (third corruption study).

Corruption has a profound effect on improving economic and social conditions, and it also affects the legislative, economic, and social conditions (Gino and Bazerman, 2009). This chapter shows that the consequences of corruption are pervasive, causing many future and sustained development problems and ultimately harming all society segments. This chapter examines corruption in light of the latest achievements and research and then presents the results (Welsh et al., 2014). Therefore, we establish a meaningful link among different theories in related research. We thus achieve norms. Because of the behavior of corruption, we must distinguish between two fundamental aspects of norms: descriptive and injunctive norms (cf. Serra and Wantchekon, 2012). Descriptive norms bring facts associated with how most human beings behave in a specified position. They also express the number of times a particular behavior has been observed. Injunctive norms bring information associated with the specific movements that almost everybody accepts or rejects – hence, whether this particular behavior is suitable and moral.

3.2 The social psychology of corruption

On the 12th of January 2010, an earthquake struck Haiti. The death toll in the field was more than 100,000. The researchers tried to find a justification for this atrocious damage (Ambraseys and Bilham, 2011). This was not the most extensive earthquake, but corruption was responsible for so many casualties. Corruption destroyed the quality of construction, infrastructure, and support structures. There is no more need to exemplify the destructive consequences of corruption around the world. Corruption is one of the most complex social issues facing different societies today (Lee-Chai and Bargh, 2001; Mauro, 1995). This destroys democracy and trust (Stiglitz, 2012). It also weakens the power of governments. It spreads injustice in societies and destroys or overuses nature (Ostrom, 2000; Rothstein, 2011a).

It is not out of the question to do a lot of research on corruption and corrupt behavior (Serra and Wantchekon, 2012) and its impact on different communities' decision-making. Laypersons and scientists like them pointed to specific behaviors in their debates about corruption, including the leader of a corporatist government embezzling public property, the exchange of gifts between government employees and citizens, and many other behaviors that can be described as 'abuses of power for personal gain' (Graycar and Smith, 2011). As will be stated, additional comments and research on corruption and corrupt behaviors will help better understand corruption and its causes from different scientific perspectives, especially when each person uses their own unique experiences and problems in research.

The current chapter attempts to evaluate psychologically corrupt activities using predictive strategies and simulate them mentally to predict future performance. Humans are uniquely equipped to experience the hedonic value beforehand (Buckner and Carroll, 2007; Gilbert, 2006), which are likely to be mainly eliminated in the future, and even events which they will in no way experience (Gilbert and Wilson, 2007). This simulation seems to be either pleasurable or painful for a practical event, in order to personally choose a certain path of movement (Schwarz and Strack, 1999).

To better address the issue of corruption, we present it as a social problem. Much of the corruption for the sake of personal preference is the moment we

consider long-term interest (Van Veldhuizen, 2013). In corruption-related studies, a power holder is challenged to choose between the abuse of power and
responsibility: corruption (abuse of power) has its interests, and no corruption
for the public's profit will be held accountable (Blau, 2009). Furthermore, to
understand the complicated dynamics concerned with corruption, we define a
distinction between personal and interpersonal corruption as we will describe.
Each of them entails unique decision-making strategies concerning the mental
illustration of expected future outcomes.

Personal corruption is a corrupt activity in which a person abuses their status
and uses it for personal gain (e.g., embezzlement and theft of public property).
Still, interpersonal corruption is where some people work together to exploit
their position for their gain (e.g., venality, kick-back payments, and setting up
fixed networks). Given these differences, the psychological components of corruption are identified, and the expected consequences of individual and interpersonal corruption occur (see Figure 3.1).

To illustrate the unique distinction between personal and interpersonal corruption, the following examples are necessary: consider a government manager
with the power of budget management training a special aspect of behavior. The
administrator discovers a misconception in the accounting system that results
from an inaccurate reporting of the actual budget of education and indicates that
the administrator intends to pocket the education budget. Here the manager has
a social dilemma and embezzles money instead of allocating it to schools. This
type of private corruption increases probably if the administrator anticipates low
chances of formal reprimand and low psychological costs such as the feeling of
guilt towards the victims of that corrupt activity, i.e., the targeted recipient of the
education funds.

On the other hand, for interpersonal corruption, the government director
needs an accountant's assistance to transfer public property. In that case, every
person is interested in participating in corruption (Weisel and Shalvi, 2015). If
the accountant makes a corrupt offer, the manager will face a complex social challenge. He is involved in deciding between his wishes, an accountant's suggestion,
and the collective interest, a decision that can have many consequences. Here, the
accountant's probable conduct affects the cost/profit estimation (e.g., the constant risk of disclosure). Besides, a manager has to make a tricky decision; in fact,
he has an ethical choice between fairness, 'Be fair and give money to the public,'
and loyalty 'Be faithful and act for the profit of your corrupt player' (Dungan
et al., 2014). If the latter outstrips the former, management expects that there
will be no guilt for interpersonal corruption. Thus, as will be elaborated in more
detail, the interpersonal corruption problems are more significant than the personal corruption problems because they are all directly involved in corruption.

In other words, personal and interpersonal corruption refers to particular psychological aspects of decision-making. Although there are differences between
crime and corruption, there is also no theoretical framework for the root (Clinard
and Quinney, 1973; Finney and Lesieur, 1982) and nature of this approach (e.g.
Amundsen, 1999; Heidenheimer, 1970; Pinto et al., 2008). This lack is mainly

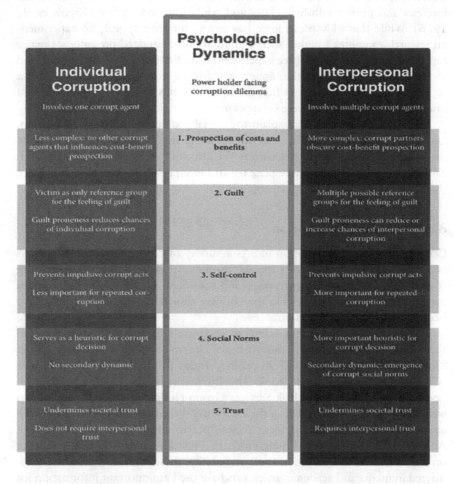

Individual Corruption	Psychological Dynamics	Interpersonal Corruption
	Power holder facing corruption dilemma	
Involves one corrupt agent		Involves multiple corrupt agents
Less complex: no other corrupt agents that influences cost-benefit prospection	1. Prospection of costs and benefits	More complex: corrupt partners obscure cost-benefit prospection
Victim as only reference group for the feeling of guilt	2. Guilt	Multiple possible reference groups for the feeling of guilt
Guilt proneness reduces chances of individual corruption		Guilt proneness can reduce or increase chances of interpersonal corruption
Prevents impulsive corrupt acts	3. Self-control	Prevents impulsive corrupt acts
Less important for repeated corruption		More important for repeated corruption
Serves as a heuristic for corrupt decision	4. Social Norms	More important heuristic for corrupt decision
No secondary dynamic		Secondary dynamic: emergence of corrupt social norms
Undermines societal trust	5. Trust	Undermines societal trust
Does not require interpersonal trust		Requires interpersonal trust

Figure 3.1 Illustration of the intra- and interpersonal dynamics of corruption and how they differ for individual and interpersonal corruption. Source: Köbis, NC (2018).

due to the neglect of corruption in personality, morality, and social psychology. Widespread social communication and multilateral decision-making processes in the field of corruption make it a confusing matter.

The road to bribery and corruption: slippery slope or steep cliff? There are numerous, almost daily, media reports about new corruption cases in various contexts, such as banking, sports, or politics. These scandals raise questions about how severe corruption emerges. Like most popular media, many scientists suggest that severe ethical transgressions such as corruption unfold gradually, a process that is frequently referred to as a 'slippery slope' (Ashforth and Anand, 2003; Bandura, 1999; Darley, 2005; Festinger and Carlsmith, 1959; Gino and

Bazerman, 2009). The belief is that power holders progressively neglect others' interests and pursue selfish interests and 'slide into' corruption (Kipnis et al., 1976). While this widespread logic has strong intuitive appeal, no experimental research examines whether such a gradual process indeed presupposes major forms of corruption. With three experimental studies using a recently developed methodology, we examine the slippery slope metaphor's validity and contrast it to a cliff metaphor that argues that corruption comes about by people seizing a one-time opportunity of severe corruption.

In the following, we first explain the social problems caused by corruption. Using the framework of this social dilemma, we study corruption and describe the common psychological mechanism. We also study the five major psychological factors related to corruption, personal and interpersonal corruption, regarding the many differences in potential cognition.

3.3 Evaluating corruption

Correctly evaluating corruption inflicts a variety of consequences. The analysis of corruption generally faces a specific hurdle, especially if corruption is more covert; also, measuring the corruption and the course of corruption is a complex methodological issue. Surveys in the country show that empirical studies have been carried out on corruption.

Currently, three special assessments and one ethical index have attracted the attention of the public and students: International Transparency Ethics Understanding International Transparency Index (CPI), Ethics Control Index (CCI). It is called the World Bank. The Inequality Index (CI) has been developed through the Political Risk Services Group (Judge et al., 2011, p. 95ff) as well as the Heritage Foundation for Economic Freedom (EFI) index. All indicators examine fixed measurement techniques to allow real-time comparison. Anomaly levels are measured using more than one source with complete reliance on expert opinions. Because of their availability, these indicators are often considered in media and academic articles and are used as important information for economic analysis (Ko and Samajdar, 2010, p. 508).

Since the beginning of 1995, the CPI has been projected through Transparency International, a Berlin-based NGO. The index is primarily based on surveys by experts, journalists, and company executives and is quite common in research as a credible and reliable provider of ethics measurement. By comparison, since its inception in 1996, the World Bank's CCI has emphasized the importance of good governance and ethical ideals as playing a potential role in the poverty alleviation mission. Besides, the CCI is considered valid, indicating a correlation of 0.97 with the CPI. Such overlap indicates that the CCI statement is significantly in line with the CPI. In line with this, the CCI includes specific indicators that measure, among other things, the ability of a business to influence policy strategies and the abundance of monetary payments in the country concerned. In 1980, CCI was a cumulative indicator of perception, primarily based on estimates by experts in the participating country, indicating a correlation of 0.75 with CPI (Judge et al., 2011, p. 95f). Finally, in 1995, economic freedom was evaluated in more than 179 countries because of the EFI's failure. Countries move from 0 (suppressed) to 100 (free) with 10 economic freedom measures such as regulatory efficiency, the openness of markets, and corruption. In 2012, the global average of economic estimates in last year's assessment dropped from 59.7 to 59.5. Despite widespread aspirations, the rating is at its highest level since 2008 at 60.2 and is rising (Heritage Foundation, 2012, p. 1). This index provides a manageable approximation of fertile land where corruption is spread. In general, it can be stated that the indicators all suffer from equal defects. Since corruption is not directly measurable, indicators include corrupt beliefs and perceptions and encompass all kinds of prejudice. From this point of view, the significance of subsequent moral tactics is at least debatable.

By presenting unique methodological strategies to combat the impediment to accurate, ethical measurement, the second task generates a timing shift in corruption measurement. Researchers have already begun to come up with a variety of indicators for measuring corruption. Based on the corruption that has gained the widespread attention over the past decades, research shows that everything in this context and the real-time dimensions have a significant impact on perceiving the corruption. This means that the effects of corruption are strongly influenced by the importance of time in corruption research (Judge et al., 2011). Hence, it is a fundamental fact that is considered when comparing the time series of corruption.

While global research has provided significant evidence of economic development corruption, the path of causality seems problematic, and simple regression does not provide any causal link. On the other hand, as an alternative study, it shows the correlation of unknown origin. As pointed out by Rose-Ackerman (2006), it is difficult to maintain the causes and consequences of corruption since causality seems to be always moving in every direction. These issues mainly arise from later problems since the commonly used indicators are based on alternate comprehension rather than 'real' corrupt analysis. Relying on fundamental errors, introducing instrumental variables (IV), and fractional evaluation can be useful. Therefore, one wants to replace the endogenous variable using IV connectors (cf. Alesina et al., 2003). The researchers attempted to solve the endogenous problem by separating a dominant correlation between the independent variables and the error term by using the introduction of a new variable that ultimately affects the primary (e.g., corrupt) neutrality variable and dependent variable (e.g., GDP) (Azfar et al., 2001, p. 51). Instead, the discovery of variables that meet the prerequisites for endogenous testing is challenging.

Furthermore, perception-based metrics are likely to be biased because, for example, foreign companies may not have the burden of paying more money than domestic companies pay. Micro-level analyses gained importance in recent years and provided a solid basis for discovering ethical factors and consequences. As a result, these strategies must be coupled with regional and laboratory experiments to conduct empirical investigations of corruption.

Lambsdorf (2012), along with these lines of comparison, argues that the principle of rational choice is accompanied by two seemingly contradictory results, one with and one without the legality of the current barriers. On the one hand, barriers have to investigate inaccuracies more than ever because this is due to the fact – at least in the absence of scales, values, and the like – that rational calculations completely drive fraudulent behaviors. 'But contrary to this account, we find heads of states who serve their people, public servants who obey the law, people who profit out of existence, and residents who died in non-combat situations. We risk morality; we put it' (Lambsdorff, 2012, p. 280).

On the other hand, because injustice is not the best Nash equilibrium subset, its actual occurrence may already be surprising. As usually happens in specific procedural settings, popularity plays no role, indicating that bribees have no incentive to respond to venal behaviors. Hence, the venal predicts the bribee's deviant behavior (for example, pocket money except for transferring related services). As a result, he will not have to pay in the first place. Even in duplicate settings, the exchange will eventually end, for what is referred to as a game effect shows that the bribe is diverted at a point of interdependence. This implies that, by using back induction, mites also avoid paying proximity. Calculating these seeming contradictions shows that 'Rational calculus, its calculation of reprimand, is only half the answer to explain human decision making. What is semi-specific?' (Lambsdorff, 2012, p. 280).

The other half is likely to respond by using a behavioral approach. Therefore, it must be borne in mind that human decision-making is the most likely of all DMs rather than the characteristics of the underlying environment, institutions, scales, values, and the like. Identifying the individual is essential as long as they interact with any given environment and therefore influence the man or woman (Akerlof and Kranton, 2000).

As a result, a full appreciation of personal decision-making in the techniques of criminal literature calls for an inclusive approach. Improving satisfaction with the sustainability of limited rationality and using exploration as an alternative to human ability to utilize the complete computation of costs and benefits to maximize utility at any time and the benefits of behavioral perspective (cf. Gigerenzer/Selten, 2002; Schmid, 2004; Wacker and Roberto, 2010; Cartwright, 2011). The body of developing techniques is in addition to replacing the logical priority method, the substitution. Subsequent chapters provide evidence for personal decision-making concerning ordinary and corrupt crimes, particularly beyond conditions of clear, rational calculation of costs and benefits.

3.4 Nash equilibrium and corruption

3.4.1 Cost-benefit evaluation between the players

In venal behavior, what is the final result and bribees exchange is their interest. The venal desires to achieve advantages which by some distance outweigh the cost, and the bribee is willing to exchange for the bribee's 'gift' for their power via providing venal things and advantages which results maybe cannot get in ordinary conditions; then, this is the typical venality pattern.

In this decision-making process, money-makers want to be fully aware of possible police decisions because they prefer to impact their gaining. Likewise, bribees also fully decide on the conceivable decision of monetary decision-makers when choosing. Uniquely, both maximize their profits by thinking about others' decisions. Here we analyze the above approach with static play. There are two players, namely venal and bribee. They both have two pure methods that are anomalous, and the super method's blends lead to a generous payout. To ensure their earnings, we first need to test their costs and benefits, which is the most critical condition for gaining balance.

State the two players as A (bribee) and B (venal). Without losing generality, consider that the two players better understand each other (that is, a fully-fledged game). Player A gains the power and interests that Player B expects from A. Player A gains such benefits using anonymity. Player B can earn more than the competition. However, in addition to the dangerous costs, it also causes the insects to be risky and causes them to behave.

Here d means the cost, which includes the cost of currency, the cost of ethics, and the cost of risk. Of course, if Player B doesn't invalidate the game with the help of credible competition, the different types of 'cost' would be zero, and at the same time, his additional income would be zero. However, there are other anxious situations where Player B chooses money, while Player A no longer chooses the bribe. Under such circumstances, Player B would also be willing to pay a specific cost, such as e.

For Player A, when Player B pays him, he weighs his interest and incomes from psychological warfare before he decides to approve. Here he has two choices: get Polio or refuse to pay for Polio. Suppose he refuses to speak up and maintains his honesty. In that case, he will gain ethical pride in attention and psychological security, from now on even stop a fascinating story, so we calculate under this scenario that he will pay B. Then he replaces the public power for a positive amount of malice and announces his repayment as A. Given the current system and related psychology knowledge, we believe that the cost of the risk and the moral cost he has paid benefits from the avoidance of guilt. Based on the above evaluation, we can describe the game with the matrix presented in Table 3.1. Each pair of numeric values in the brackets represents the payout symbols A and B, respectively.

Table 3.1 The game between bribers and bribes

B \ A	Bribery	No bribery
Bribery	$(c - d, a - b)$	$(-e, b)$
No bribery	$(0, 0)$	$(0, 0)$

3.4.2 Nash theory evaluation

As we all know, each of the players follows profit maximization. Due to this fact, count on and were $a, b, c, d \succ 0$ and $a \succ 2b, c \succ d$ for the players' incomes and costs. The game system is that players, by thinking about different player strategies, choose the optimal strategy that gets the most out of them. Here is a work on the dominant strategy technique for Nash alignment. Through the above payback matrix, we see that for Player A, there is no photo to choose Player B;, his payout of from the 'bribe' is generally not much less than the bribe, mostly $a - b \succ b$, $0 = 0$. Therefore, in the simple strategy set of Player A, 'bribe' is superior to 'no venality.'

Similarly, for Player B, 'bribe' is a strictly dominant strategy because of $c - d \succ 0$, $0 = 0$ the pure strategy. Then by turning off the dominant strategy approach, it is useful to gain a particular focus on game theory, namely '(absurdity, numbness).' We can also determine that the Nash equilibrium '(numbness, numbness)' is also the optimal Pareto solution. Therefore, players are more likely to identify anomalies. The above analysis shows 'throwing a long line to catch a big fish,' 'If you can't give up your baby, you'll never meet a colorful wolf.'

3.4.3 The Nash equilibrium between venals

In real life, many monetary recipients also cannot accept this kind of behavior in their hearts and cannot be forced to perform unlawful behaviors by competitors and managers or supervisors usually with the help of these types of unhealthy behaviors. They may also be concerned that the opposition will become unfair because of the competitors' reality. Then, after doing the same thing, they will have nothing to regret. Here we are thinking about the static game with full information, and there are only two players, players A and B. Each player has two pure strategies, injustice and injustice. Also, the repayment matrix is shown in Table 3.2, where a_i, b_i, c_i, d_i are all negative ($i = 1, 2$). From the assessment of the cost and benefit, we are aware that, according to the opposition's strategy, players' profits from distrust are not always less than zero, so we have $a_i \geq c_i$ and $b_i \geq d_i (i = 1, 2)$. Moreover, it is convenient to get the game's unique Nash equilibrium point with the described approach, that is '(venality, venality).'

With the condition that Players A and B are upsides (including venality numbers), there is an implied condition in the payoff matrix: $d_i \geq a_i (i = 1, 2)$. When each side chooses 'no venality' and competes fairly, each of their predicted utilities (i.e., payoff values) are equal, specifically $d_1 = d_2$; if each chooses venality,

Table 3.2 The game among bribe-givers

A \ B	Bribery	No bribery
Bribery	(a_1, a_2)	(b_1, c_2)
No bribery	(c_1, b_2)	(d_1, d_2)

then their payoff values are the margin of the predicted utility and the quantity of venality, respectively. So, there will be $d_i \geq a_i$ ($i = 1, 2$). At the moment, the game seems to be a 'prisoners' dilemma.' The optimal Pareto game solution is not Nash's unique equilibrium point but the '(no phobia, numbness)' situation. The above debate suggested that a 'self-care strategy' might be helpful. However, this requires your players to treat others as you wish to be treated, but only if they do the same. The Chinese also said, "Do unto others as you would not have them do unto you."

3.4.4 The evaluation of the venal's motivation and the reasons for corruption risk

The above analysis has shown us this revelation: Nash's theory of venality play is entirely at the players' expense and wages. The reason why the eliminating-dominated strategy method can be used is that we gave the hypothesis $a \succ 2b$ and $c \succ d$ showing that each player's profits are a lot higher than their cost in a successful venality process. If we change their repayment in the following ways: improving the risk of expenses and ethical recipient expenses:

(1) Promoting powers of reward for the absence of punitive and supervisory measures for malicious conduct, on the other hand, changes in the cost of flanking, and if they are inconsistent with it, even without really ugly accusations, they will be punished accordingly.

Therefore, under such a policy, the game's payoff matrix can be re-written as Table 3.3. Now we evaluate the Nash theory for the new game. We assume that at the element of Nash equilibrium, Player A's optimal strategy is $(p, 1 - p)$. The optimal strategy of Player B is $(q, 1 - q)$, i.e., Player A chooses the strategy 'bribe' with the probability p and chooses the strategy 'no venality' with $1 - p$; similarly, Player B chooses his strategy 'bribe' with q and the strategy 'no venality' with $1 - q$.

Therefore, Player A intends to choose a reasonable probability for optimizing the following parametric programming (P_A):

$$\max \ pd(a - b) + (1 - p)qb - bp(1 - q) \tag{3.1}$$

Table 3.3 The new game of bribers and bribes

B A	Bribery	No bribery
Bribery	$(c - d, a - b)$	$(-e, b)$
No bribery	$(0, -b)$	$(0, 0)$

Thus, we achieve the Nash theory for the new game as a mixed strategy situation $((p^*, 1 - p^*), (q^*, 1 - q^*))$.

$$q^* = \frac{b}{b - a} \tag{3.2}$$

Here, as we can see, when the parameter e improves, the optimal probability of p^* improves.

$$\max pd(c - d) - eq(1 - p) \tag{3.3}$$

Moreover, the optional solution is:

$$p^* = \frac{e}{c - d + e} \tag{3.4}$$

Thus, we achieve the Nash theory for the new game as a mixed strategy situation $((p^*, 1 - p^*), (q^*, 1 - q^*))$. Here as we can see, when the parameter e improves, the optimal probability p^* improves. This means that because Player B's costs increase, the likelihood of a 'no pay' reduces. By evaluating the above, we find some profound reasons. In real life, although innocent people suffer moral condemnation and legal reprimand, the practices of utility continue. The aim lies in its inherent economic factors and its incentives to monitor and reward and punishment for whole system.

On the one hand, it is difficult to distinguish between legitimate and illegitimate exchanges in social interaction clearly, and interpersonal differences are challenging to account for because of the reality of their compatibility. On the other hand, everyone hopes to expand their interests. In the venality game players are all rational to make the most of their tools. Although we all condemn such behavior, if there is such a need, everybody will come up with such a reason and act. Given this fact, in addition to guiding people's attitudes toward the 'personal' and 'social optimum' approaches, improving reward and punitive system behaviors are among the required methods available. When it comes to paying a good amount of profit, nobody has the incentive to pay.

3.5 Development of theories and conceptual theories

The magnitude of the interaction between goals and actual behavior depends on three conditions: first, it depends on how specific the behavioral intention and criterion analysis is. It also depends on the degree of consistency of the goals concerning the difference in measurement time and performance. Finally, the extent to which a person deliberately plays a role is crucial (Madden et al., 1992, p. 4). Experimenting with this method, meta-analyses indicate successful model predictions of behavioral goals and actual behaviors useful for identifying goal strategies for behavior change (Sheppard et al., 1988).

The theory of planned behavior introduces the third aspect by extending the previously defined simple voluntary control boundary conditions, which affect behavioral goals: realized behavioral control. After reasoning (Ajzen, 1991, 1980, 1985), realized behavioral control is the main issue with having (realized) the resources and opportunities to perform a particular behavior and exhibiting an extrinsic effect on both (indirect) behavioral intentions and actual (direct) behavior.

> The indirect impact is based on the assumption that realized behavioral control has motivational implications for behavioral intentions. When human beings believe that they have little control over performing the behavior due to the fact of lack of requisite resources, then their intentions to operate the behavior may be low, likewise if they have favorable attitudes and subjective scales concerning the performance of the behavior.
>
> (Madden et al., 1992, p. 4)

Both strategies are appropriate for explaining corrupt behaviors due to psychological factors, not just economically rational behavior. Here the idea is conveyed that real behavior results from cognitive processes, where attitudes, subjective scales, and personal beliefs about their control over particular behaviors play a decisive role. Experimental evidence supports the notion that personal self-esteem is moderately influenced by the ability to perform a particular action (for example, accepting a bribe). It predicts actual behavior (for practical aspects cf. Bandura et al. (1980)).

As Eliasberg (1951, pp. 326–329) points out, from a psychological perspective, one's actual behavior is unlikely to be the result of calculating the correct trade since this risk has not been adequately analyzed. Instead, 'corruption is a psychologically unsatisfactory and incomplete, legally wrongful and motivationally unethical alternative for public sympathy.' Psychological effects that make a person's behavior corrupt are diverse, developing interdependencies between the inner and outer world, which are explained in Figure 3.2.

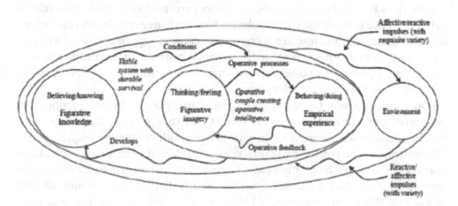

Figure 3.2 Impacts on behavior from a psychological point of view. Source: Yolles (2009, p. 703).

3.6 Nash theory and the first study on corruption

3.6.1 Data and materials for analysis

Throughout the study, we examined whether realized descriptive scales were correlated with corrupt behavior. To this end, we evaluated the perceived abundance of this particular corrupt behavior with an earlier case of corrupt play (described in more detail below).

Participants. Students in Polish society who had graduated from programs of higher education from 'Adam Mickiewicz University, Gdansk University of Technology, Jagiellonian University, Kozminski University, University of Lodz, University of Social Sciences and Humanities, University of Warsaw, University of Wroclaw,' including eight different Polish universities and including entrepreneurs who had graduated from programs of higher education and other entrepreneurs who had not received any training in ethics ($N = 66$, Mage = 26.79, SDage = 15.49; 51.5% = female) took part in the study in exchange for direction credit or money (€2). Participants first answered several questions assessing the realized scales associated with workplace-related behavior – one question assessing the particular corrupt practice modeled in the ensuing corruption game.

Ethics statement. All of the studies mentioned that players use the same set of basic experiments. Our College Ethical Analysis Board (VCWE) is the author of this pilot series. In all studies mentioned in this chapter, participants completed a written consent form before completing the scale. Upon completion of the study, participants were happy and thanked for their participation. In all of the studies presented, we assessed participants' age, sex, and education level before recreation. These demographic elements had no statistically significant effect on corrupt decision-making in any reported studies (all $ps > 0.122$).

Prior scales. The workplace scale assesses work-related behavior with five objects ($\alpha = 0.724$; e.g., 'Copy a company-owned software program for your use'). Players indicated the perceived abundance of the described behavior on a 100-point slider answer scale ranging from '0' (= nobody does it) to '100' (= anyone does it). Higher scores indicate higher perceived abundance than related behavior. Since none of the current objects have evaluated specific corrupt behavior in the game, and because of the corruption context's specificity, we developed a new one to evaluate scales related to corrupt behavior in a corrupt game. It says, 'Invite a government official to take a personal vacation at the expense of the company to secure business advantages.'

Corruption game. Corrupt play requires three players. In one auction, two players compete for a full prize of 120 credits. The third player offers the award to the best bidder (see Figure 3.3). In each round, both competing players receive 50 credits from the bidder. Rival players can be selected from a range of options that are not available from other bids (0 credits) to invite their entire budget for rounding (50 credits). Contestants submit credits they do not currently allocate in one bid. While the best possible bidder wins the overall prize, the award will be split equally between the two bidders if they present the same. The bidding process continues for five rounds.

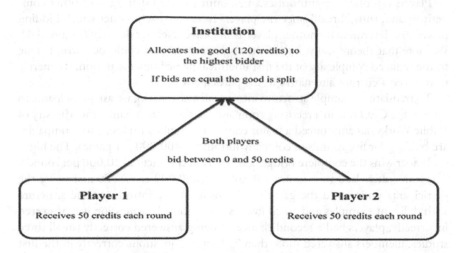

Figure 3.3 The triadic structure of the corruption game in which players take the role of the potentially corrupt participants.

Table 3.4 Two players' outcome matrix of the fair bidding game

Player 2						
50	**40**	**30**	**20**	**10**		

Player 1	50	40	30	20	10	
50	60 60	120 10	120 20	120 30	120 40	120 50
40	10 120	70 70	130 20	130 30	130 40	130 50
30	20 120	20 130	80 80	140 30	140 40	140 50
20	30 120	30 130	30 140	90 90	150 40	150 50
10	40 120	40 130	40 140	40 150	100 100	160 50
0	50 120	50 130	50 140	50 150	50 160	50 50

The payback matrix (see Table 3.4) shows all the potential effects of this bidding process. Allocating 50 credits to the bid is the dominant approach to this bidding process – these alternatives are in the exact Nash 2 balance. In other words, for each proposed player of 50 credits, bidding solution achieves excellent results independently of the other player's choice. We are made up of a corrupt option for a player in this free bidding structure. Our strategy is similar to the usual triple structure for many corrupt trading deals: two (or more) competing players – one potential corrupt player and one fair participant (i.e., a victim of corruption) – and one player. Third, it looks like an official sacrificing to the best possible bidder. We only report all players in this analysis to the performance of a potentially corrupt player.

Note: the matrix shows the results for each player earlier to introduce the corrupt option into the game. There are different bid options for each player in the different listings. The players affect each other. The dominant strategy for each player is 50 credits.

Players can offer the authorities an amount to avoid sharing with other competitors and, thus, 'break' since the players have told each bidder which bidding player can determine if another player is negative. Neither do they (Figure 3.4). We note that theoretically, any of the competing players could be corrupt. Due to the reduced complexity of the first unethical game implementation, we merely introduced a corrupt alternative to the participant.

To translate this simple structure into a realistic scenario, we ask participants to assume the CEO role in a recovery company. In the current game, the Ministry of Public Works has announced a major contract to build a bridge. Two companies are bidding for five rounds of company funding ($400,000 per game). The highest bidder wins the complete bridge construction contract ($120,000 per round). The same offers lead to a contract shrinkage ($60,000 each). To make sure the participants understand the game's bidding form, we introduced the structure with different examples and asked five test questions. If a question was answered incorrectly, players had a second chance of being answered correctly (in all three studies, members answered more than 72% of the questions correctly in the first session).

This trend is common in commercial transactions but can be considered corrupt because it guarantees personal profit to the minister and leads to a bidding profit for the player (Heidenheimer et al., 1989). Due to its standard functioning and legality (for example, lobbying practices), we refer to this choice as 'corrupt obscurity' in this section.

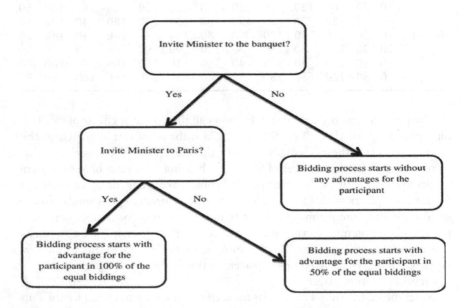

Figure 3.4 Third corruption game was used in the first study in which players make a step-wise decision associated with this issue whether to bribe the minister or not.

There is a second invitation opportunity for those who celebration on the minister, including inviting the minister's to take a private holiday on the company's budget. This Tender guarantees the benefits of 100% of the same. The second decision implies more severe and illegal conduct because the company budget is used to provide the minister with a particular benefit in exchange for the full benefits of the bridge project's dedication. We call this choice 'grossly corrupt' because of its illegal nature.

3.6.2 Reported results from the first study

The present study's main purpose is to examine the positively theorized link between perceived abundance and actual involvement in a particular corrupt behavior (see Table 3.5 for an overview of corrupt behavior). To this end, we performed two binary logistic regressions. In the first regression, an a priori corrupt object was used as a predictor and it was decided to propose a banquet (without bidding versus bidding) as a dependent variable. The results show that the perceived abundance of corrupt activity significantly influenced the minister's decision to ban ($B = 0.77$, Wald = 5.08, Exp $(B) = 1.86$, $p = 0.024$). The increase in the perceived abundance of widespread corruption has increased the minister's chances of being invited to the banquet, considering 1.86 amount of the factors.

Note: Table 3.5 shows the number of players selected for the invitation or avoiding the invitation for each occasion. Note that the only player who invited the minister to the banquet faces a second decision not to leave the minister.

In the second binary logistic regression, we used the same predictor and used vacation choice as a deployable variable (invitations or no invitations to holidays). Again, we find a significant effect ($B = 0.64$, Wald = 3.85, Exp $(B) = 1.89$, $p = 0.05$). An increase in the perceived abundance of a standard deviation from the corruption improved the minister's chances of being invited to leave based on 1.89 amount of factor. The more a player is corrupt in both types of corruption, the more likely they are to participate. Importantly, the workplace scale did not predict any of the dependent variables (all ps > 0.236).

3.6.3 Discussion associated with the result of the first study

The results support our hypothesis and show that which scales are related to the corrupt behaviors. Suppose the players thought that inviting the minister to take

Table 3.5 Overview of the player's decisions in the first study

Did participants invite the minister?		
	Yes	*No*
First decision (bidding to the banquet)	42	24
Second decision (bidding to vacation)	22	20

a private vacation on the company's budget for access to jobs was relatively common. In that case, they might also be involved in this form of corruption. These realized scales also predicted the likelihood of being attracted to a variety of more severe and more ambiguous anomalies – the minister's invitation to the banquet.

This is probably due to the substantial similarity between the two types of invitations. The fact that only the particular unethical case – and not the whole workplace scale – predicts corrupt behaviors again reflects corruption's contextual characteristic.

To support an alternative explanation for the results, it is likely that responding to the belief-related question would increase the happiness scales'; to rule out this possibility, we conducted a second study (Bicchieri and Xiao, 2009). This time, we measured realized scales after the game of corruption so that the analysis of corrupt scales would not influence the decision to conduct corrupt behaviors.

3.7 Nash theory and the second study on corruption

3.7.1 Data and materials for the analysis

During the second study, we further simplified the game of corruption by removing the structure of two dependent variable steps. We follow such step-by-step approaches to modeling many corrupt events that follow a slippery slope strategy (Darley, 2005). Therefore, only the player involved in obscure corruption decided whether or not to pursue more severe corruption. This decision-making structure deprives a large part of the sample of the second corrupt decision. To divide the interpretation of the findings and clearly show the relationship between realized scales and the severity of irregularities, we eliminated the ambiguous non-ethical option. So players were directly faced with the choice of whether they would like to pay for a minister's private vacations (i.e., more corrupt corruption) and win 100% of the bids (see Figure 3.5 for a game of three simple corrupt games).

Participants and protocol. Students in Polish society who had graduated from programs of higher education from 'Adam Mickiewicz University, Gdansk University of Technology, Jagiellonian University, Kozminski University, University of Lodz, University of Social Sciences and Humanities, University of Warsaw, University of Wroclaw,' including eight different Polish universities and including entrepreneurs who had graduated from programs of higher education and other entrepreneurs who had not received any training in ethics ($N = 119$, Mage = 21.57, SDage = 2.80, 63% = female) participated for course credit or money (€2). Participants first played the unsuitable simple game and then demonstrated their realized descriptive scales. Apart from simplifying the dependent variables, the game was used in the same way as in the first study.

Post hoc scales. Upon completion of the corrupt game, all participants responded to a question that assessed their perceived abundance of holiday bidding (i.e., 'How many people do you think would recommend a vacation minister?') six – Scale from '1' (= no one) to '6' (= all). Participants could

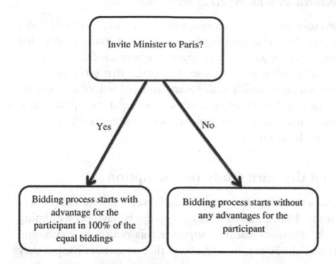

Figure 3.5 Game three of the simplified corruption game in which the players at once face the choice of whether or not to bribe the minister with the vacation.

choose four different percentages to the abundance (1–25%; 26–50%; 51–75%; 75–99%). We also manipulated public awareness of the three conditions. The players were either in a cube that turned on a webcam, a webcam was off, or there was no webcam at all. This manipulation did not affect the reported results.

3.7.2 Reported results from the second study

In the second study, we assessed whether the descriptive scales (measured after corruption play) relate to corrupt behaviors. We computed a binary logistic regression with realized scales as a predictor and a corrupt decision (i.e., holiday invitations) as the dependent variable. We searched for a significant effect ($B =$ 1.42, Wald = 26.72, Exp (B) = 4.162, p and it $^\varsigma.001$). Improving realized scales of corruption through a standard deviation has increased the minister's chances of being invited to take a vacation based on factor 4.16.

Given the importance of early influences for ripe picking (Bargh and Chartrand, 1999), we also examined whether filling in the WPN scale in the first study caused unethical behavior. We tested whether the level of anomalies previously in the first study was higher than in the second study, where participants did not complete the WPN scale before the corrupt game. We did not see any difference between holiday tenders (χ^2 = 0.119, p = 0.730) or in the ambiguous corruption in the first study and severe corruption in the second study; hence the first choice both studies (χ^2 = 1.16, p = 0.28).

3.7.3 Discussion associated with the result of the second study

The results again show a sure link between corrupt behavior and descriptive scales of this form of corruption. In the second study, players were asked to describe their scales after playing a corrupt game. Participants may perceive their scales to change according to their behavior in the game (i.e., scales that serve as rationales). Alongside the first study, in which scales were assessed before moral play, we strongly support the close link between descriptive scales and corrupt behavior in play. The more participants perceive corrupt behavior, the more likely it is to corrupt behavior in an unethical game.

3.8 Nash theory and the third study on corruption

One way to interpret the scale difference obtained in the second step is in the second logical strategies study. Individuals may justify corrupt behaviors by claiming that this is a 'joint work.' Previous research supports this notion by examining corrupt behavior (Mazar et al., 2008a, b) and shows that people are lying if they have more excuses or explanations. Descriptive scales may be an excuse (Kerr et al., 1997) and a common technique. Ethical constructions may vary, meaning that individuals may change their realized personal scales in their behavior (Ashforth and Anand, 2003). We argue that in addition to acting as an explanation of corrupt behaviors, descriptive scales also provide a precedent for corrupt behavior.

To test this assumption and provide causal evidence, we conducted a 0.33 study in which we tested whether manipulating descriptive scales could influence corrupt behaviors. Previous research suggests that small associated ethics can reduce unethical behavior such as fraud (Mazar et al., 2008b; Bryan et al., 2012). Using descriptive scales related (Wenzel, 2004) to primes, a set of behaviors ranging from improved tax compliance to enhanced energy consumption behaviors is presented (Nolan et al., 2008; Schultz et al., 2007). However, there is no empirical test of the effect of creative descriptive numbers on corruption. To investigate the cause-and-effect relationship of realized scales with corrupt behavior, we conducted an experiment where we manipulated descriptive scales by providing participants with a brief introduction before they made an unethical decision.

3.8.1 Data and materials for the analysis

We used the same simple study design as the second study. In addition, we reduced tenders from five to one to reduce the complexity of computing tools for participants. Before that, the players had to anticipate the corrupt advantage of five bids. Now they just have to wait for unrealistic profits to get out of the bidding. Economically, ethical profit is the same, but it is easier for participants to calculate corrupt profit.

Participants and protocol. We conducted an online study ($N = 259$; Mage = 35.65; SD age = 11.54; 42.1% = female) in English through Amazon Mechanical

Turk. Participants were eager to reside in the United States and accepted more than 5,000 HITs with a minimum agreement rate of 98%. Participants participated with a one dollar refund. Participants first read the unethical game instructions. They received one of three creative statements before deciding whether to invite the minister on vacation. In the context of anti-corrupt scales, we make our players read an immediate statement, 'Almost no one invites the minister.' In favor of the scale, fast reads, 'Almost everyone invites the minister.' In the control condition, the participant received the message, 'It's not that urgent.' After completing the corrupt game, we re-evaluated the realized descriptive scales with the same question as assessed in the second study. Also, half of the players received time pressure notifications, which did not affect the reported results.

3.8.2 Reported results from the third study

Measurement of realized descriptive scales. To investigate whether scale manipulation influenced realized scales of corruption, we conducted an ANOVA with scale manipulation (three levels) as a predictor and examined realized scales (post-game). We examine post-game scale as a dependent variable. Results showed significant differences in realized scales between the three groups ($R = 0.34$, $F(2.225) = 70.9$, $p < 0.01$). Participants in terms of corruption scales perceive corruption as the control condition ($M = 4.03$, $SD = 1.25$), which considering corruption at the end, to be the least common. Is ($M = 2.57$, $SD = 1.33$). Post-game was more common than participants in the pro-scale condition ($M = 4.74$, $SD = 1.06$). All post-holiday micro-group comparisons are significant (all ps < 0.002; modified Bonferroni).

Hypotheses testing. In the first step, compared with the other two studies, we find that the total corruption level in the online study was higher than the two pre-test laboratories ($\chi^2 = 9.37$, $p = 0.009$), most likely due to an increase in anonymity. We then tested whether scale manipulation by computing logistic regression represented scale manipulation as a predictor variable and the decision to engage in corruption as a dependent variable. In this third study, participants in the ethical scales condition were significantly fewer than the participants in the control condition of irresponsible scales ($B = 0.83$, $B = 6.43$, 0.6, $p = 0.011$, prediction (2.30)). The odds of being attracted to corruption in the anti-corrupt scale were 2.3 times lower than in the control condition.

In addition, we found a significant difference between the condition of anti-corruption scales and the condition of pro-corruption scales ($B = 0.79$, Wald $= 5.87$, $p = 0.015$, Exp $(B) = 2.30$). The odds of being attracted to corruption were 2.3 times higher than those of anti-ethical scales. There was no significant difference between control and pro-scale conditions ($p = 0.89$). Thus, the first anti-corruption scale has significantly reduced the level of corrupt behaviors in play compared to the corruption scale's control.

Besides, we assessed whether realized scales play a role in influencing manipulated scales. Two mediation analyses using bootstrap evaluate two significant effects (anti-control; anti-scales) indicating complete mediation in each case (anti-control: CI95% [–1.83; –0.73]; anti-pro-scales: CI95% [–2.59; –1.2]).

3.8.3 Discussion related to the results from the third study

The results of the third study explain that descriptive scales can influence subsequent levels of corrupt behaviors. When players receive a urgency message indicating a low abundance of corrupt behaviors, the extent of the maladjustment is greatly reduced compared to control conditions and pro-corrupt scales. Interestingly, the results show no difference in corrupt behavior between the pro-scale and the control condition, which indicates that individual corrupt behavior is generally considered universal.

3.9 Theory base

Broadbent and Laughlin (1999) identify a substantial research agenda relating specifically to private finance initiative (PFI), although most of the issues they identify would be relevant to many kinds of asset-based 3Ps. Most of this plan relates to the public policy and economic impact of 3Ps on government. However, they stress the importance of analyzing the role and decision-making processes of the private sector.

While the tendering of 3Ps contracts is required in northwestern Europe procurement law, the benefit of using competitive tendering is illustrated both by practitioners' experience and academic literature. Substantial savings may be achieved by the competitive tendering of services formerly provided by the public sector (McAfee and McMillan, 1988; Deakin et al., 1997; Domberger, 1998; Kavanagh and Parker, 2000). However, quantifying the benefits of contracting is not very 'simple' (Deakin and Walsh, 1996) or 'soft' (Lane, 1999) is more problematic. Auction theory scholars have demonstrated an optimal level of competition and, thus, it is possible to have too much as well as too little. If there are too few bidders, value for money is not achieved; if there are too many, and this is generally known, potentially strong bidders may be reluctant to bid as the chances of winning are too small (McAfee and McMillan, 1986). It has also been contended that strong bidders bidding within a field containing weaker bidders may bid a higher price to carry out a service than they would otherwise to avoid the 'winner's curse,' i.e., winning a contract at an unrealistically low price (Han and Shum, 2002). While the effect of bidder numbers on competition has been demonstrated, academic research on bidding, which is primarily based on mathematically derived auction theory, tells us little about why these outcomes occur (Maskin and Riley, 2000; Klemperer, 2004).

In their discussion of competition in bidding for government contracts, McAfee and McMillan (1988) make three contentions that reflect the central tenets of auction theory. First, they show that competition drives down prices and, therefore, the government's cost of purchasing goods and services. Second, they suggest a relationship between the contracting process's institutional structure and the response of bidders. The former comprises the proposed contract form (e.g., fixed price, cost plus, or incentive-based) and the auction (i.e., competitive procurement) process. Third, they contend that both the choice of institutional structure and the tendering process's results will be influenced by the risk orientation of both government and potential contractors. Auction theory is used extensively (and successfully) to inform auction design (for example, mobile communications licenses – see Klemperer, 2004; Milgrom, 2004). However, there are arguably problems in operationalizing the conclusions of auction theory where the selection of a contractor is on quality and price, an issue identified as requiring further research (Klemperer, 2004).

Furthermore, risk orientation, in the context of 3Ps bidding, has multiple meanings. In the sense used in auction theory, risk orientation relates to a bidder's willingness to accept a lower surplus to enhance the probability of winning the contract (Klemperer, 2004; Milgrom, 2004). This is quite a different meaning from the concept of risk as typically used in 3Ps (and, indeed, in other forms of contracting), i.e., the acceptance by one of the contracting parties of the costs of uncertainty in a specific area (McAfee and McMillan, 1988). Both are important, but the existence of two meanings suggests that bidders make two different risk assessments. The first assessment is of the prospect of winning; the second is of the long-term costs of delivering the project within a payment level judged to be acceptable to the commissioner. The latter may be known; in 3Ps procurement, the 'affordability envelope' (i.e., the public-sector budget for the project) is often, but not always, made known to bidders.

It should also be noted that the literature varies in the assumptions on the risk orientation of both the public and private sectors. McAfee and McMillan (1988) assume for analysis that governments are risk-neutral while recognizing, in practice, that they 'sometimes exhibit an aversion to risk' (p. 14). Walsh (1995) describes the public sector as 'highly risk-averse' (p. 122) and suggests this explains the preference for fixed-price contracts. The common assumptions on risk orientation are arguably insufficient to cater to quality-based contracts like 3Pss and do not sufficiently distinguish risk orientation concerning the bidding process (i.e., for the private sector; for the public sector, the likelihood of awarding a contract).

The literature indicates that the design of a bidding process is influenced by the commissioner's knowledge or perceptions of the market, not merely by the project's nature. Walsh (1995) points out, in the context of the tendering of social care services, that local authorities write tougher contracts in countries in northwestern Europe where there are more potential providers. Therefore, the implication is that bidders can be more selective the more potential contracts there are on the market, or where there are fewer competitors or both.

Transaction cost economics (TCE) has frequently provided scholars with an analytical framework for the study of public/private contracting. TCE contends that firms seek to minimize transaction costs when procuring assets or services. However, the more specialized the asset is required, the more the purchaser and supplier are at risk of each other's opportunistic behavior. Each makes investments that cannot easily be deployed elsewhere. Hence, highly specialized assets need to be produced through a governance structure that limits the scope for opportunistic behavior.

While Williamson, and much of the empirical research using TCE, focuses on the transaction as the unit of analysis, Williamson posits a three-level schema for analyzing transactions, comprising the institutional environment, the governance model, and the actions of individuals (Williamson, 1996). In this schema, 'the institutional environment is treated as the locus of shift parameters, changes in which shift the comparative costs of governance' (p. 223). But he acknowledges that TCE has concentrated too intensely on the transaction and too little on the institutional environment.

This is an important issue concerning the public-, as opposed to private-sector contracting. In the private sector, Williamson's contention that there will be an optimal governance structure related to asset specificity can be given practical effect; a firm may choose, in the light of the nature of the asset it requires, to 'make or buy.' As Vincent-Jones (1997) points out, in the public sector, not only is the mode of governance mandated, but so is the specific contractual form, describing the system as 'imposed and regulated contracting' (p. 144). It should be noted that the government of the country in northwestern Europe strongly resists the suggestion that 3Ps is mandated and argues that the choice of procurement route is purely a value-for-money judgment (HM Treasury, 2003). However, scholars have challenged this assertion (Grout, 1997; Gaffney et al., 1999; Froud and Shaoul, 2001; Shaoul, 2005).

The literature on trust forms an essential element of the theory base on public contracting, although scholars differ on the nature and role of trust in transactions. Numerous authors explore the broader role of trust within the development of public/private contracting (e.g., Coulson, 1997, 1998; Deakin and Michie, 1997). Davis et al. (1998) describe trust as a mechanism for reducing uncertainty and complexity in trading relations. Ghosal and Moran (1996) argue that the development of trust within trading relationships is a substitute for formal contractual mechanisms. Indeed, formal mechanisms (by suggesting a lack of trust) encourage opportunistic behavior. Conversely, Deakin and Michie (1997) and Poppo and Zenger (2002) argue that the development of trust and formal contractual controls are complementary, and Deakin and Michie emphasize how institutional forms (including the form of contract) sustain and reproduce trust.

Arino et al. (2001) stress the importance of the 'initial condition' informing successful alliances, the information and preconceptions that the parties have about each other before they start negotiating. It has been suggested that where trust can be created or reinforced (for example, through the form of contract), contracting out even highly specialized requirements may be optimal (Chiles and McMackin, 1996). This is because if the parties trust each other, the risk of opportunistic behavior lessens. The trust literature recognizes that levels of trust are self-reinforcing, being increased or decreased based on the parties' experience. Given the scope for repeated transactions, trading relationships are commonly described in game-theoretic terms, with continuing relationships having the characteristic of a repeated game with the resultant tendency against opportunism and towards cooperation (Axelrod, 1984). However, empirical studies primarily focus on trading relationships and collaborations once established rather than during the bidding process. The 3Ps procurement process is often not a repeated game because of public-sector personnel changes, discussed in the last section. To that extent, and because contractor selection is not solely price-based, it may be seen by bidders as a game with ambiguous payoffs, making conventional game-theoretic analysis more difficult to apply (Hargreaves Heap and Varoufakis, 2004).

Thus, the theory base with which to analyze 3Pss is well developed but focuses more on the relationship after contracts are signed than on the pre-contract and bidding process. However, theory, and most particularly auction theory, suggests that pre-tender procurement design, and the perceptions and risk orientation of the commissioner and bidders, are critical to understanding the bidding process.

3.10 Conclusion and expert judgment

Contrary to the widespread belief that corruption comes about through a slippery slope, the present studies provide novel evidence suggesting that the path towards severe corruption might instead be a cliff. Across three studies, people were more likely to engage in severe corruption when this option was presented abruptly rather than gradually, even though they did acknowledge the unethicality of severe corruption. The moral evaluation of severe corruption and the (combined) economic costs and benefits did not differ across the different conditions.

Given that most scientists, and laypeople alike, believe in the slippery slope analogy, it is essential to ask the obvious: how can we account for evidence favoring the cliff rather than the slippery slope? One line of reasoning is that the intuitively compelling notion that repeated transgressions lower moral thresholds may not always be accurate. Rather than a process of habituation and moral disengagement, people may seek to avoid repetition of corruption because it is expected to be psychologically taxing – mostly when the corrupt opportunities occur in short succession (Mazar et al., 2008b). It poses another threat to one's self-image, and therefore even a second more minor form of corruption can be undesirable (Study 3).

When deciding whether to engage in unethical behavior, people take both the external costs and benefits and the psychological costs and benefits of the respective act into account (Messick and Bazerman, 1996). Unlike previous studies (Welsh et al., 2014), we kept the economic costs and benefits constant across different conditions. Thus, our findings point towards a new psychological factor – the sequence of decisions. A single severe act, directly presented to the participants, might be easier to justify, causing less tension between being a moral person on the one hand and enjoying the benefits of dishonesty on the other hand (Batson, 2016).

A complementary argument is that a single act requires less intentionality and planning than repeated behaviors (Batson and Powell, 2003). The enormous benefits might reinforce a selective focus on self-interest rather than a positive self-image. In contrast to previous work (Welsh et al., 2014), this study looks at bribery, a form of unethical behavior that entails a collaboration between multiple corrupt agents (Köbis et al., 2016; Weisel and Shalvi, 2015). The resulting local social utility ('I also help the other,' Ayal and Gino, 2011) might give rise to reputational concerns regarding the other agents, which then further facilitate these self-serving justifications.

Future research is needed to examine the underlying mechanisms and boundary conditions. For example, how severe corruption emerges under varying punishment regimes requires future research. Yet, given the ubiquity of the slippery slope's belief, we conclude with two lessons from the present research. One lesson is that people may be more willing to engage in severe, single (and perhaps unexpected) instances of corruption than widely believed – even if they recognize the corruption of these behaviors. Another lesson is that repeated forms of unethical behavior may be more psychologically taxing than most of us tend to believe, especially if the second occasion brings about smaller benefits for one's self. These findings thus shed light on an unexplored area of sequential corrupt decision-making. Overall, our findings suggest that those willing to engage in bribery seek to obtain the most significant advantage for the lowest moral price. Instead of going through a process of repeatedly engaging in corruption (slippery slope), they are rather opportunistic once (cliff).

The rest of this research is as follows. First, we concentrate on theoretical arguments and refine testable hypotheses. We then describe the source of the data, variables, methods, and findings. Finally, we conclude with a discussion and outcome for policymakers and firms. Scholars Luis Enrique Orozco, Jaime Páez, Germán Montoya, and Javier Pombo all accepted that although one cannot generalize, some business owners participate in corrupt acts when we considering players age, education, or the age of their businesses. Some entrepreneurs place economic interests above social interests, but scholars focused on the following:

I.   Neither the effect of business owners on corrupt practices nor the extent to which corruption pervades their thoughts and actions has yet been analyzed.
II. Entrepreneurs have not been profiled concerning their ethical behavior.
III. Entrepreneurs and their performance respond to rules that emerge from the environment.
IV. The environment is polluted because values are not strongly imbued either in the home environment or schools.
V.   Entrepreneurs are not born corrupt. Over time, they approve the idea of corruption as a feature of scalable behavior.
VI. What did universities stop doing that used to produce good young students? When they graduate as professionals now, they are a symbol of national corruption.
VII. Personal and professional success is measured by the possession of things and the desire to enhance income more quickly and with less effort. This thinking may convince business owners to engage in unethical practices to achieve this version of success.

VIII. Education must be modernized as is being done in the business world, and it must incorporate ethical topics more effectively than is the case in traditional programs.

IX. Numerous scandals in Poland, including corrupt contracting in urban improvement (carousel de contratación, Los Nule), price-fixing for consumer products (el cartel de Los pañales, el cartel de Los Cuadernos), and corrupt investment schemes (Interbolsa), make it clear that there is a crisis of values in the business environment.

X. University graduates are very well prepared for entrepreneurship but lack a grounding in business ethics.

XI. Universities have a responsibility for their graduates' business ethics, but the role of the family, society, and the social environment must also be considered.

XII. Jaime Páez indicates that between some of his students who have started businesses, their entrepreneurship's early years were characterized by a frenzy to establish the company, innovate, and act ethically. Still, with time their ethical training faded from memory, their ethical method waned, and they adopted unethical practices prevalent in the business world.

The effect of descriptive scales on corrupt behaviors has often been theorized, but as far as we know, it has never been tested (Rothstein, 2000; Bicchieri and Rovelli, 1995). The current set of studies provides the first empirical support for the hypothesized effect.

Perceived descriptive scales were associated with subsequent corrupt behavior (the first study of corruption). To rule out that creative health enhances this effect, we introduced a realized scale difference in addition to evaluating scales after evaluating corrupt behavior (the second study of corruption). Finally, concise statements containing creative descriptive data have been successfully affected. In particular, information showing a low abundance of corrupt behaviors (ethical scales) has reduced the level of subsequent corrupt behaviors (the third study of ethics). Anti-corruption scales may have affected the perceived abundance of corruption in the relatively large sample.

People trust descriptive scales as a guideline for making corrupt decisions, especially in situations in which they have little or no personal experience (Cialdini et al., 1990; Hogg et al., 2008). We place participants in such uncertain circumstances which are descriptive scales, primarily based on beliefs and not on our personal experience. Participants in the study, who rely heavily on scales, suggest that people who are confronted with such particular conditions may be particularly susceptible to descriptive scales in particular selection criteria.

Experience in organizations that do not have personal experience with business: if they believe that corruption is not commonplace, they will likely. Our results provide a new experimental path for how scale-related reminders can form corrupt scales. Small reminders and messages can make a 'dumb' (Thaler and Sunstein, 2008). Curb corruption – especially in literature where people lack first-hand experience or mistakenly believe that a high percentage are involved in corruption.

From a methodological point of view, corrupt play provides an indirect test tool that allows scientists to realize psychology that contains corrupt factors.

Also, it accepts corrupt test decisions when participants have firsthand experiences, even based on which they report an ethical dilemma. Covering up corruption in a bidding game allows for the study of corruption while avoiding favorable social effects. Using stupid, corrupt play, we forecast the first definition of the link between realized and corrupt descriptive scales.

It is noteworthy that no financial incentive or reprimands in the three studies were presented in this chapter. Considerations of tangible results or cost-benefit calculations are unlikely to reflect current findings. Unlike most ethical investigations in which reward and reprimand play an essential role, we opted for this design because it allows us to identify descriptive scales as the primary driver of uncovering an isolated environment's corrupt complexity. Indeed, one crucial issue for future research lies in evaluating how creative influences in all literature types, including those that are incentives and reprimands, strike corruption. Future research can also conclude that descriptive scales influence the realized reprimand for corruption.

Besides, does the perceived abundance of corruption increase when the potential profit is high? Although the relationship between the abundance of reprimand and descriptive scales may seem plausible, what is the relationship between reprimand severity and descriptive scales? This question is fascinating in gray areas of ethics, such as the ambiguous morality raised in the first study of corruption.

In the third study of corruption, we show that scales are influenced externally, for it causes the short-term to successfully influence the realized descriptive scales. On a positive note, we found that anti-corruption scales reduced the level of subsequent corrupt behaviors. Since we have used relatively specific manipulations of scales in the third study, the use of implicit manipulations is a more exciting way of investigating corruption, especially since many forms of abuse of power occur outside of conscious awareness (Lee-Chai and Bargh, 2001).

Recent related research shows that the belief that corruption is spreading affects the likelihood of corruption. To obtain generalizable results, we could put participants in a corrupt position. Still, due to some limitations, we asked them to imagine a corruption situation that would impact the results as well. Given the personality of corruption, studying corruption in real-life contexts poses a significant challenge to corruption research. Corruption transactions and operations, as stated in this study, provide a way to study corruption empirically. Further studies can examine the impact of descriptive scales on other forms of corrupt behavior to generalize the results obtained, in other ways, such as economic factors (Dungan et al., 2014) and remuneration and fines (Henrich et al., 2010).

Many studies are being carried out to discover how corruption scales are shaped. For example, will media reports about corruption affect corrupt behavior? How? Previous research has indicated that media coverage is strongly influenced by descriptive scales, which leads to imitation of advertising behaviors (Cialdini, 2006; Phillips, 1974). Whether a similar 'copycat' corrupt behavior follows a well-known illegal act (for example, the Madoff case) may be an exciting topic for future research.

Small-scale or micro-corruption occurs in small dimensions within the social frameworks and customs. For example, exchanging inappropriate gifts or using personal communication for profit (another instance uses the word rather than the exchange of gifts (favoritism) in favor of family, friends or specific groups, etc. leader and the organization, government or military. The question often asked of those kinds of people is: **why are you doing this? You realize that this is to the detriment of some and for the benefit of the person or persons you are targeting.** But the answer is more interesting. **If I don't, who does?** In dictatorial or growing societies, such cases are often abundant). This type of corruption is particularly evident in developing countries where public servants live well below the poverty line. Large-scale or macro-corruption occurs at the highest government levels and requires the overthrow of the political, legislative, and economic system. Corruption generally occurs in dictatorial governments and governments that lack the proper laws to prevent corruption. The system of government in many countries, including the branches, legislative, executive and judicial, which operate independently and under the rule of law, they must keep away from corruption. Systematic corruption (or endemic corruption) is corruption that occurs primarily because of a weakness in an organization or process. It can be seen as opposed to corruption, where only a few employees or agents of an organization are involved in corruption. Promoters of systemic corruption include conflicting interests, arbitrary power, monopoly power, lack of transparency, low wage pay, and a culture of impunity for corruption.

We reported the importance of more accurate psychological indicators for descriptive scales realized in corrupt behaviors as the first empirical support. The notion that corruption is widespread has a significant influence on corrupt behaviors and decision-making – perceptions that small-scale pressure can influence. Alster described social scales as 'the cement of society' because of the importance of descriptive scales in everyday decisions. In an entirely corrupt society, the 'cement of social scales' is conducive to corruption. Still, in a society with less corruption, the opposite is the case, and corruption is eliminated so that things can go to other scales (Elster, 1989). Referring to the first example, a series of recent related studies show that your answer to the question of whether you would bribe the minister is related to your belief in the corrupt behavior of others. The idea that no one gives the proposed minister reduces the amount of corrupt activity. At the same time, if you consider it as a general business activity, the areas of corruption will increase for you, and you will say: 'I invite the minister, who does not?'

3.11 Conclusions and further developments

The literature suggests that the critical factors in understanding private-sector bidding decisions relate, first, to the public sector's choice of governance structure for the transaction, second, to the level of trust between the transacting parties, and, third, to the influence of the respective institutional environments of the transacting parties on mutual preconceptions and the development of trust. As noted in this chapter. 'The government largely mandates theory base,' the governance structure for 3Ps (although changes after the fieldwork – such as the batching of individual schools and hospital PFI projects into larger contracts – suggest that greater diversity may exist in the future). Trust, the institutional environment's influence, and the significance of developing relationships to be a successful bidder emerged as important issues. From the bidder's perspective, all these processes are part of a risk-management process over which the bidder wishes to maximize control and influence.

Reflecting on the analysis of the interview evidence and the literature review, an initial model has been developed of how the private sector views a potential PFI/3Ps project. This is shown in Figure 3.6.

In line with the literature and empirical evidence, it characterizes the decision as comprising two assessments, one the perceived risk of bidding, the other the perceived risk of the project. The factors that determine the 'risk of bidding' are twofold: the first element is an assessment of the probability of winning, this being influenced by the availability of credible partners and the opportunity cost of bidding (i.e., alternative use of the resources). The other element of the assessment concerns the level of trust in the commissioner's competence and commitment. It should be noted that the empirical evidence did not reveal that either the number or identity of potential competitors bidding for 3Ps contracts influenced the decision. This may be explained by the interviewees being well established in their respective sectors of the 3Ps market. However, auction theory would suggest that the perceived competition does influence bidder behavior.

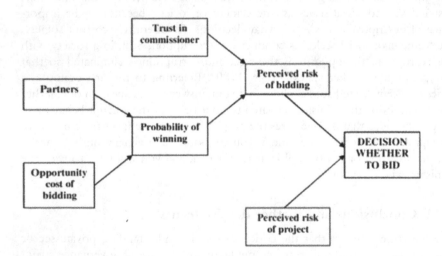

Figure 3.6 Factors influencing private-sector bidding for 3Ps contracts.

Despite the procurement problems identified by interviewees, all indicated that their firms intended to stay in the 3Ps market. This suggests that some factors must outweigh others that weigh heavily against bidding. This research sought to identify what the factors may be but is purely exploratory. Further research is needed to test the robustness and completeness of the factors, identify the relative importance of each, and explore whether and how the relative importance changes. For example, is the nature of the firm more significant than the perceived competence of the commissioner? For example, will a firm with plenty of scope for deploying its resources outside 3Ps be more influenced by perceptions of the commissioner than a firm dedicated to 3Ps? A better understanding of the influences on bidders' risk orientation (in particular, the extent to which the actions of the public sector influence this) could make a valuable policy and theoretical contribution.

There are also implications for government policy and practice. The government has put a great deal of effort into standardizing PFI contract documentation to lower transaction costs, and it sees this as a means of increasing competition:

> It is in the long-term interest of the public sector that there are easy access and fair competition. To that end, we are keen to ensure that transaction costs are reduced because we believe them to be higher than they need be.
> (Interview with the Head of HM
> Treasury's PFI Unit)

However, a financial adviser interviewed questioned the relative importance of standardization as a policy issue: 'Standardization is generally good but certain types of the scheme (particularly the bigger infrastructure projects), will never easily lend themselves to Standardization. The real issue is to have a sufficient deal flow to keep the market buoyant.'

Williamson's examples of the factors that constitute the institutional environment are fundamental structures such as contract law. What is less developed within the literature is what might be termed 'lower-level' factors that fall within Williamson's definition cited above in the 'Theory base' section (Williamson, 1996, p. 223). This research suggests that the factors that determine whether bidders bid for 3Ps contracts may usefully be explored by developing a model of the institutional factors that influence bidder decision-making, an issue not substantially addressed in the literature.

As Williamson notes, TCE has focused on the transaction and not the other two levels. Within 3Ps, as the mode of transaction is virtually mandated, developing an understanding of the other two levels' significance is important for both policymakers and practitioners. Furthermore, Nooteboom (2004) has contended that the central unit of analysis within TCE, the transaction, is better described as the 'transaction relation,' as the tenets of TCE (e.g., opportunism) are characteristics of a relationship rather than a transaction. This research would suggest that this theoretical refinement is of value as it allows a useful distinction between 3Ps as a mandated policy of government and 3Ps as a transacting relationship that demonstrably works to the greater satisfaction of the parties in some areas than others.

There is plenty of literature about corruption. Unfortunately, most articles have not been published in project management journals and, more importantly, are not focused on the project management community. Despite the relevance of corruption in public projects and megaprojects, it seems that it is not convenient to research, talk, and write about it. Despite the number of papers published in relatively close fields (governance, stakeholders, risk, value), corruption seems taboo, and project management scholars seem afraid to write about it. However (and unfortunately), the context of the public sector and procurement of large projects is ideal for corruption.

This chapter adopts the institutional theory to introduce a 'corrupt project context' to indicate a systemic view's research needs. Therefore, a major contribution of this chapter is to rethink the role of corruption in projects from a social and institutional level. Projects are not delivered in a vacuum, but several internal and external stakeholders link them to the project context. Simultaneously, the study of the project context alone is not entirely appropriate since different project characteristics may favor or disfavor corruption. The investigation of the successful planning and delivery of megaprojects, which have most of the characteristics that can favor corruption, in a corrupt project context is, therefore, a critical challenge that practitioners and academics will need to face in the next decades. If project management practitioners and academics aim to reduce corruption, it is not enough to point out a 'few rotten apples' when an entire area or country suffers endemic corruption. It is unrealistic and simplistic to say 'don't deliver projects in that location,' primarily due to the urge for projects and megaprojects in the next decades, such as power plants and hospitals (to name a few). So, there is a massive scope in the project management sector to research how to deliver successful projects in a corrupt project context. Maybe, in power plants, corruption has to be fought through the delivery of several small standardized projects (with an established cost-benefit track record) rather than in megaprojects with unique characteristics and budgets challenging to estimate.

The case-study approach used in this chapter has some limitations, as it cannot quantitatively demonstrate exactly the specific impact of a corrupt context on a project's poor performance. As discussed in the chapter, demonstrating such a relationship is remarkably difficult because corruption is often considered a phenomenon more associated with individuals than with projects. Furthermore, megaprojects often have unique characteristics and many influencing factors for their performance. Therefore, the chapter provides a first tentative approach to the problem, whereas further research will need to identify the most appropriate methods to analyze the phenomenon in depth. If the debate is to be moved forward, a better understanding of corruption from the project management perspective needs to be developed.

First, future research activities should deepen the correlation between corruption and project performances, comparing the costs of similar megaprojects in countries with different levels of corruption. The biggest challenge is to isolate corruption from other factors that may lead megaprojects to failure.

Second, future studies should investigate the correlation between project management attributes (e.g., contract forms, risk allocation strategies, etc.) and corruption and then investigate the causation behind the correlation.

Third, it is necessary to develop tools and guidelines, expanding the work presented in (IPMA, 2015), a short, ten-page document with a Code of Ethics and Professional Conduct (actually, the relevant section is three pages). Even if the document does not explicitly mention corruption, it acknowledges its existence. It states that: 'Whenever possible, we avoid real or perceived conflicts of interest, and disclose them to affected parties when they do exist ... We reject all forms of bribery' (p. 6). So, this code of conduct is a starting point, even if it lacks an action plan.

Fourth, corruption does not merely lead to extra cost and delay but also increases transaction costs, such as the effort to set a specific procurement and controlling system. For instance, Zhang et al. (2015), looking at land hoarding in China, found that when the net loss from corruption, 'income from corruption minus the penalties for corruption and cost of strict inspections,' is less than the cost of strict inspections, the final evolutionary stable strategy of the inspectors is to carry out insufficient inspections. The topic of increased project transaction costs is vastly under-investigated.

Lastly, it is necessary to develop tools and control systems to address a project's 'corruption performance.' Corruption in megaprojects is most likely the main cause of their inefficiencies, and the academic community should not deliberately ignore this aspect in the project management literature. **There is an elephant in the room; let's talk about it!**

Note

1 The present chapter's results are significantly connected with the PhD dissertation of Mohammad Heydari, which was written at the Nanjing University of Science and Technology, entitled: 'A Cognitive Basis Perceived Corruption and Attitudes Towards Entrepreneurial Intention.' Supervisor: Professor Zhou Xiaohu, School of Economics and Management, Nanjing University of Science and Technology, Nanjing, Jiangsu, China. For more information about this dissertation, you can contact Mohammad_Heydari@njust.ed u.cn and njustzxh@njust.edu.cn. There are some questions contained in this chapter which suggest a purpose for further research. Also, it is necessary to mention that this chapter is the result of the ten years of research in different countries on 'human and organizational behavior.'

4 Decision-making criteria for selecting and evaluating Public–Private Partnerships

Case studies

4.1 Introduction

Public–Private Partnerships (3Ps) have been widely applied across the world by governments to provide a series of critical public services (e.g., health, education, water, and electric power supply, and transport) since the global financial crisis, owing to the limited funds available for main infrastructure development. Despite more and more successful operations of 3Ps, some project failures have still been reported in the literature (e.g., cost overruns, schedule overruns, and stakeholder dissatisfaction). There are numerous factors critical to a 3Ps's success, and practical performance evaluation maintains a decisive role (Liu et al., 2014c). According to Yong (2010), the debate about 3Ps has moved from the ideological argument of their advantages and disadvantages to the management to structure them well to achieve the predetermined goal. Therefore performance evaluation must be adequately addressed during 3Ps delivery.

There is widespread consensus that evaluation is critical to business success, regardless of whether it is at the corporate or project level (Love and Holt, 2000; Kagioglou et al., 2001), Liu et al. (2014b) maintain that an ineffective and incomprehensive evaluation can act as a trigger for producing below-optimum service quality of infrastructure. Nonetheless, most 3Ps projects have not undergone a complete form of evaluation in terms of what has been delivered (Hodge, 2005). Most 3Ps evaluations solely concentrate on meeting budget and construction schedules (Liu et al., 2014d). The 3Ps markets in Australia and the United Kingdom, for example, are considered to be sophisticated and mature (Hodge, 2004); however, ineffective and incomprehensive evaluation was identified as a significant factor that had negatively affected the delivery of Latrobe Regional Hospital, Deer Park Women's Prison, and Ashfield Prison (Victorian Audit-General's Office, 2002; House of Commons, 2003; Roth, 2004).

A 3Ps's success is essentially determined by the project's construction and various issues identified in the project's pre-design and pre-construction stages (Liu et al., 2014a). Yet, the evaluation for the vital issues in the early stages of 3Ps, which is usually referred to as *ex-ante* evaluation, has surprisingly received limited attention. Against this contextual backdrop, this chapter reviews the normative literature and conceptually proposes a new framework for 3Ps *ex-ante* evaluation.

DOI: 10.4324/9781003177258-4

Since the worldwide financial crisis, limited funds have been available for major infrastructure projects. Many governments and other agencies have had great difficulty promoting and financing infrastructure development using Public–Private Partnerships. Therefore, it is crucial that supporting agencies and financiers conduct effective pre-design and pre-construction analysis and evaluation of the proposed projects. Traditional theories have tended to focus on meeting budget and project duration targets. These single-dimension evaluations may not account for the more complicated legal, organizational, and financial interfaces created by multiple-stakeholder interactions. More effective approaches, mechanisms, and frameworks are crucial in this early-stage volatile environment. This chapter attempts to address some of the vital issues involved in the inception stages of major infrastructure projects using 3Ps. Stakeholders must have a useful performance measurement framework to evaluate 3Ps before making crucial decisions on major infrastructure projects.

Fundamental to creating a new approach is understanding what evaluation methods are used by reviewing 3Ps' critical literature. The chapter concentrates on 3Ps characteristics and performance measurement. Analysis of the context demonstrates that a more feasible framework based on the Performance Prism can capture more of 3Ps' distinct characteristics. It is a more suitable framework for conducting these early project evaluations, providing more efficient performance measurement.

This chapter can provide essential guidance for designing and selecting the most effective performance measurement indicators for 3Ps in infrastructure projects.

3Ps are the long-term contractual relationships formed between public and private entities, aiming to procure and provide public assets and relevant services through the use of private-sector resources and expertise (European Investment Bank – EIB, 2004). Governments, essentially, embrace 3Ps as they offer the following benefits: (1) accelerated infrastructure provision through allowing the public party to translate capital expenditure into a flow of on-going service payments (income); (2) timely project implementation through the allocation of design and construction responsibility to the private sector; (3) reduced whole-life cost and motivating performance offered by the strong incentives of the private sector to minimize costs and improve management over a project's life-cycle; (4) reduced government risk exposure by transferring such risks to the private sector; (5) improved service quality and innovation through the use of private-sector expertise and performance incentives; and (6) enhanced prudent management of public expenditure and reduced corruption by the increase in accountability and transparency (European Commission, 2003, p. 15).

Compared with other forms of private participation in infrastructure, 3Ps' defining features include risk transfer, long-term contract relationships, and partnership agreements (Akintoye et al., 2003). Kwak et al. (2009) state that 'the complexity of contractual relationships between participants, and the long concession periods associated with 3Ps, makes them distinct from a traditional infrastructure development routes' (p. 56). The exploration of the normative literature of 3Ps suggests that the previous studies on 3Ps primarily concentrate on six areas, including critical success factors, the public sector's roles and responsibilities, concessionaire selection, risk management (e.g., identification, analysis, and allocation), project evaluation, and project finance (Liu et al., 2014b).

4.2 The effect of 3Ps and Multi-Criteria Decision Making Process

With the ever-increasing pace of urbanization in China, the limited financial sources of the government have been unable to keep up (Liu et al., 2016; Ameyaw and Chan, 2016; Song et al., 2018; Tao et al., 2018). To relieve infrastructure funding pressure and raise the rate of funding, an increasing number of local governments in China began to encourage the private sector to participate in public investment projects. Therefore, 3Ps have been explored and become a preferred solution for current infrastructure projects in the public sector (Carbonara et al., 2014; Wibowo and Alfen, 2015). To be more specific, a 3Ps project is long-term cooperation between the public and private sectors for the joint construction of urban infrastructure projects or offering public goods and services based on a concession agreement (Zou et al., 2014). In 3Ps infrastructure projects, the formation of partnerships is a type of Multi-criteria Decision Making between the public and private sectors. 'Sharing interests and responsibilities and long-term cooperation' are the most critical principles of 3Ps infrastructure projects (Chou and Leatemia, 2016). During the construction and operation process, the cooperative partners need to share interests and responsibilities within the public and private sectors. Thus, the appropriate selection of cooperative partners in a 3Ps project is the basis of the project's success (Ng et al., 2012).

However, it is often difficult or impossible to give decision-making subjects (i.e., administration organizations) a precise evaluation of matching objects with several possible evaluative values. Moreover, group decision-making is used most in modern decision-making techniques to try to avoid mistaken decision caused by cognitive bias, and thus there might exist disagreement among the decision-makers (DMs) who cannot persuade each other (Ai et al., 2016). Taking buying an established house as an example, suppose that four members in the buyer's family are asked to describe the degree to which a house is superior to another. The members prefer to use values between 0 and 1 to express their preferences. One member provides 0.65, one provides 0.75, one provides 0.75, and the other provides 0.80. To describe the original evaluative information, obfuscation tools, such as a fuzzy set, or intuitive fuzzy set, have their limitations. As a generalized form of fuzzy set, hesitant fuzzy set (HFS) (Torra, 2010), whose membership degree is allowed to be several evaluative values, provides new ideas for solving hesitation and disagreement of DMs. In the above example, the degrees to which a house is superior to another can be represented by a hesitant fuzzy element (HFE) {0.65,0.75,0.75,0.80}. Besides, determining the evaluative criterion weights is also important since it significantly affects the matching decision's accuracy (Wang et al., 2016).

To overcome the above limitations, this chapter introduces hesitant fuzzy set into Multi-criteria Decision Making and considers the multi-criterion decision of 3Ps infrastructure projects, in which the information of evaluative criterion weights is completely unknown. Specifically, the objectives of this study are (1) to determine evaluative criterion weights objectively through maximizing the deviation between evaluative criteria and constructing an optimization model; and (2) to obtain an optimal matching scheme by calculating the compatibility of the two parties via Hesitant Fuzzy Element Ordered Weighted Averaging (HFEOWA), which is further used to construct an optimal matching model. Also, the method proposed in this chapter can be further expanded the gap in the hesitant fuzzy value.

4.3 Evaluative Infrastructure based on 3Ps

A 3Ps is generally considered a win-win mechanism that represents a collection of resources and efforts for both the public and private sectors (Zhang et al., 2016). Multi-criteria Decision Making of 3Ps infrastructure projects involves both the evaluation of the enterprises by the government and the evaluation of the government by the enterprises, which results in a long-term and stable partnership, thereby ultimately maximizing the satisfaction of both sides (Chou et al., 2016; Ismail, 2013). Therefore, both the government and the enterprises' matching satisfaction degree should be fully considered in 3Ps infrastructure projects' Multi-criteria Decision Making process (Kwak et al., 2009).

Selecting private partners is a crucial issue for 3Ps infrastructure projects. Rai et al.,(2005) identified evaluative criteria that are classified into financial, technical, safety, health, and environmental, and managerial for 3Ps projects in general (Chai et al., 2013). Ng et al. (2012) indicated that it is also useful to consider the particular evaluative criteria associated with the 3Ps procurement process, such as separate financing, design, construction, and operational responsibilities. Four evaluative criterion categories of the public and private sectors for 3Ps infrastructure projects were established by Zhang et al. (2015) using a structured questionnaire survey, including finance, technology, safety and health, and the environment. In general, evaluative criteria of the public and private sectors can be classified into several categories:

- **Financial strength**: many scholars have reported various kinds of financial strength for Multi-criteria Decision Making of 3Ps infrastructure projects, such as capital strength, financing capacity (Belassi and Tukel, 1996), and financial guarantee (Aziz, 2007). This criterion is used to measure private companies' financial credibility to deal with capital crises (Hwang et al., 2016).
- **Technical ability**: this criterion category is the ability to establish project parameters and in-house expertise (Ahadzi and Bowles, 2004). The private sectors have to show their technical ability, including advanced technology, project implementation plans, personnel reserve, and critical equipment, to

carry out all construction activities required for a specific 3Ps infrastructure project.

- **Management ability**: one of the most critical factors for completing 3Ps infrastructure projects is top management ability. The private companies have to show their management capacity to carry out all construction activities required for a specific 3Ps infrastructure project, including management formalization, project management capacity, communication and cooperation, concession duration, risk management ability, and maintenance ability.
- **Relevant experience**: as a central part of the Multi-criteria Decision Making process of 3Ps infrastructure projects, relevant experience has received increasing attention in the past two decades, just as financing experience, constructing experience, and operational experience also influence Multi-criteria Decision Making.
- **Credit level**: credit hazards can easily lead to severe cost increases and schedule delays for 3Ps infrastructure projects. Thus, the credit level is crucial for Multi-criteria Decision Making.
- **Governmental insurance**: currently, there are no special laws and regulations for 3Ps infrastructure projects to define the responsibility division between participants, project assessment, contract performance and change, and conflict-solving styles. However, appropriate laws and regulations are the foundation for the private sector to participate in 3Ps infrastructure projects and the premise of multi-criteria decision making due to a long construction cycle and the increasing risk of operational project construction.
- **Governmental guarantee**: the recoverable investment rate is critical for the private sector involved in 3Ps infrastructure projects; thus, the public sector should offer a recoverable investment rate to businesses via offering a government-backed rate price, purchase, and compensation in the scope of the given benefit level.
- **Governmental credibility**: the private sector involved in 3Ps infrastructure projects must be built on project understanding, but the public sector discloses the critical information about the 3Ps project. Many previous cases have demonstrated that unclear project information leads to a lack of trust in the public sector. Moreover, the private sector is weak when playing games with the public sector; thus, the private sector's legitimate rights and interests are hard to guarantee.
- **Governmental risk-sharing**: there are a series of unpredictable risks due to 3Ps infrastructure projects characteristically lasting a long time, and it would expose the private sector to severe damage if these risks occurred (Shen et al., 2006). Therefore, the government's risk-sharing affects the private sector's quantity and quality of 3Ps infrastructure projects.

4.4 Selection Strategies

Researchers have proposed Multi-criteria Decision Making selection approaches from economic and non-economic perspectives to balance 3Ps infrastructure projects' main parts effectively. Considering the uncertainty within 3Ps infrastructure projects, Carbonara et al. (2014) established the statistical distributions of the random input variable and Monte Carlo simulation technique to calculate a '*win-win*' solution for both the project promoter and the host government in the concession phase; to allow for fair risk-sharing between the two parties to satisfy both the private sector and the government through guaranteeing the minimum interests of each party; and finally, to fairly allocate risks between parties. To evaluate and compare different 3Ps alternatives, Xie and Ng (2013) first imitated human reasoning and conducted multi-objective decision-making through Bayesian network techniques. They then proposed a decision-making approach by connecting the decision-making items, evaluative criteria, and the ultimate objectives from the government, investor, and community perspectives, respectively. A weighted score approach was used to combine the three main stakeholders' objectives into a single value. Wang et al. (2016) first constructed a bilateral matching satisfaction index system for the 3Ps project at both the government and enterprise levels. They established the matching satisfaction judgment matrices of the two sides via intuitionistic fuzzy numbers. Then, they used the *Choquet integral* to match the satisfaction evaluation vectors of the two sides and obtained the final matching results by constructing a multi-objective decision model. Ng et al. proposed a simulation model to help a public partner identify the concession period according to the expected investment and tariff regime. Then, they presented the need for establishing different scenarios to represent the risks and uncertainties involved and introduced a fuzzy multi-objective decision model to trade off the associated three concession items.

Overall, the methods mentioned above are based on general evaluating criteria and qualitative evaluating criteria weights, which are pre-assigned subjectively. However, the quality of the decision results can be evaluated by the superiority of the evaluating criteria and subjective judgment of DMs and the objective judgment of evaluating criteria weights. Thus, developing an objective way to determine the weights associated with the evaluative criteria is undoubtedly essential and exciting. Meanwhile, if there are a few different values in a membership set, traditional methods usually replace these discrete values with continuous values within an interval, which do not adequately reflect these values' differences. How to develop a proper method to handle this issue is also challenging.

4.5 Fuzzy method and Decision-Making model for selecting Infrastructure Projects

In the past several decades, the Chinese government has initiated extensive infrastructure projects to improve China's quality of life. Initially, these projects were fully funded by the government. To facilitate building more projects, the government is now inviting more private companies for co-investment and building to ensure financial sustainability. For some special projects, each private company can bid on one project due to either solemn financial commitment or limited capability and capacity. At the initial stage, the government would like to have

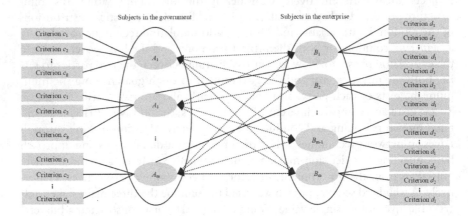

Figure 4.1 Forecasting the results.

an assessment to allocate the right projects to the right enterprises (i.e., different industries) considering their background and expertise for maximizing benefit to all the involved parties. This chapter aims to develop a unified framework for 3Ps projects, optimal matching.

During the Multi-criteria Decision Making decision-making process, the government and the enterprises are supposed to be evaluated based on different subjects. Without loss of generality, the number of issues in the government exceeds the enterprise's number of subjects. For each subject, further evaluation will be through a group of criteria. The criteria used to evaluate the government's subjects and the enterprise subjects are private and public evaluative criteria. This mutual relationship is described in Figure 4.1. From this figure, the evaluated subjects in the government are A_1, A_2,..., A_m and the evaluated subjects in the enterprise are B_1, B_2, ..., B_n; it is clear the directed dotted lines between subjects in the government and subjects in the enterprise denote government satisfaction with subjects in the enterprise and enterprise satisfaction with subjects in the government, respectively. Undirected solid lines denote the matches between the subjects in the government and the subjects in the enterprise. The lines joining the government's subjects with the evaluative criteria denote the evaluative criteria considered by the subjects in the enterprise for matching.

Similarly, the lines joining the enterprise subjects with the evaluative criteria denote the evaluative criteria considered by the government's subjects for matching. For each subject in the government, the evaluative criteria are assumed to be the same as in the enterprise. The same case is assumed for each subject in the enterprise. However, the evaluative criteria considered for the subjects in the government might be different from those considered for the subjects in the enterprise. Under this framework, the subjects in the government will measure enterprise satisfaction through the left evaluative criteria c_1, c_2, ..., c_g, and the subjects in the enterprise will measure government satisfaction through the right evaluative criteria d_1, d_2, ..., d_l.

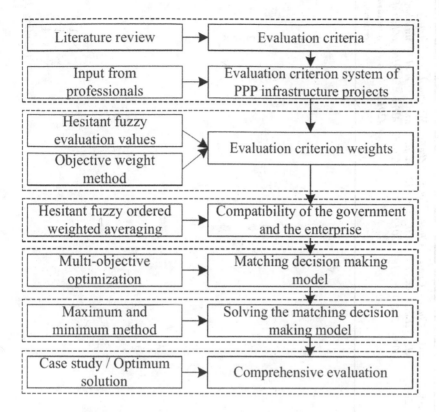

Figure 4.2 An algorithm for the research.

The research framework shown in Figure 4.2 presents processes and associated techniques used in this study. Major steps include: (1) structuring an evaluative criterion system of 3Ps infrastructure projects, (2) determining the evaluative criterion weights, (3) calculating the compatibility of both sides (i.e., the subjects in government and the subjects in the enterprise), (4) establishing the matching decision-making model, (5) solving the matching decision-making model, and (6) comprehensive evaluation.

Structuring an evaluative criterion system. Through interview and literature review, 29 evaluative criteria are classified into 9 categories, including financial strength, technical ability, management ability, relevant experience, credit level, governmental insurance, governmental guarantee, governmental credibility, and governmental risk-sharing for 3Ps infrastructure projects. The evaluative criteria of matching on both the public and private sectors under the 3Ps model are shown in Table 4.1. The private evaluative criteria and the public evaluation criteria are shown in Tables 4.2 and 4.3, respectively. All the sub-criteria will be used for the following matching decision.

Determining the evaluative criterion weights. Using subjective and objective criteria in the weight assignment method is essential for the coordination

Table 4.1 Analysis of 3Ps

Criteria	Sub-criteria	Belassi and Tukel (1996)	Zhang (2005)	Wang et al. (2007)	Aziz (2007)	Kwak et al. (2009)	Chan et al. (2010)	Chan et al. (2010)	Chou et al. (2012)	Ismail (2013)	Zhang et al. (2015)	Wibowo and Alfen (2015)	Liu et al. (2016)	Total number of hits of a certain criterion
Financial strength	Capital strength	✓		✓	✓	✓				✓				5
	Financing capacity	✓	✓		✓	✓	✓	✓					✓	7
	Financial guarantee													1
Technical ability	Advanced technology		✓	✓		✓	✓	✓	✓	✓				6
	Project implementation plans			✓									✓	5
	Personnel reserve			✓	✓		✓	✓	✓	✓				3
	Critical equipment													2
Management ability	Management formalization	✓	✓	✓				✓					✓	4
	Project management capacity	✓	✓	✓	✓		✓	✓					✓	6
	The ability of communication and cooperation	✓		✓		✓	✓	✓		✓			✓	7
	Concession duration	✓	✓	✓		✓								4
	Risk management ability		✓						✓					4
	Maintenance ability		✓						✓					3
Relevant experience	Financing experience			✓	✓	✓	✓							2
	Constructing experience	✓	✓	✓	✓									3
	Operating experience		✓	✓										2
Credit level	Enterprise qualification	✓												1
	Honor on projects												✓	1
	Social reputation				✓		✓		✓				✓	4

(*Continued*)

Table 4.1 Continued

Criteria	Sub-criteria	Belassi and Tukel (1996)	Zhang (2005)	Wang et al. (2007)	Aziz (2007)	Kwak et al. (2009)	Chan et al. (2010)	Chan et al. (2010)	Chou et al. (2012)	Ismail (2013)	Zhang et al. (2015)	Wibowo and Alfen (2015)	Liu et al. (2016)	Total number of hits of a certain criterion
Government guarantee	Legal guarantee	✓	✓	✓	✓		✓	✓	✓	✓	✓	✓	✓	11
	Approval guarantee	✓	✓	✓								✓		5
Government support	Political support	✓	✓	✓	✓	✓	✓	✓	✓	✓		✓	✓	9
	Interest rate guarantee	✓		✓		✓	✓		✓			✓	✓	8
	Price guarantee		✓	✓			✓		✓			✓		5
Government credibility	----											✓	✓	3
Government risk-sharing	Restriction of competition			✓		✓	✓	✓	✓	✓		✓	✓	7
	Information transparency		✓	✓		✓	✓					✓		4
	Contract spirits		✓			✓	✓				✓	✓	✓	7
	Risk-taking							✓	✓			✓	✓	7
		11	13	18	12	11	13	10	14	8	2	10	14	--

Table 4.2 Decomposition of evaluative standards on business subjects by government issues as a starting point

Sub-criterion	Description
1. Capital strength	Total funds, quantity, and scale invested for completed construction projects
2. Financing capacity	The scale of financing possible
3. Financial guarantee ability	The ability to obtain financing guarantee from financial institutions for the construction project
4. Advanced technology	The leading technology in the world
5. Project implementation plans	Plan formulation of activities or process of hard work from start to finish
6. Personnel reserve	Human resources to meet the demand for long-term development goal for the enterprise
7. Critical equipment	Critical equipment to ensure continuous, stable, efficient, and economic operation of the entire machinery
8. Management formalization	Degree of standardization on strategic enterprise management, marketing management, human resource management, financial and tax management, material management, etc.
9. Project management capacity	The ability to plan, organize, coordinate, implement, and control projects
10. The ability of communication and cooperation	Skills of coordination and communication with the government
11. Concession duration	Specified period of recycling project costs on investment, operation, and maintenance, and obtaining a reasonable return
12. Risk management ability	The ability to minimize the risk in a hazardous environment
13. Maintenance ability	The ability of operation and maintenance for infrastructure projects
14. Financing experience	Experiences of financing similar infrastructure projects
15. Construction experience	Experiences of constructing similar infrastructure projects
16. Operating experience	Experiences of operating similar infrastructure projects
17. Enterprise qualifications	The qualification and level of quality standard adapted to the qualification
18. Honor on projects	National or provincial honor gained by the enterprises for their completed projects
19. Social reputation	Credit evaluation is given by partners cooperating in the enterprise

and compromise of different evaluative criterion weights because the evaluative criterion weights often are entirely unknown (Kasperski and Zieli'nski, 2007). The lower difference between evaluative criterion weights means a closer preference difference between the objects and the matching objects, which is not conducive to optimal matching. Moreover, determining evaluative criterion weights

Table 4.3 Credit appraisal is done by partners who work together in the company

Sub-criterion	Description
1. Legal guarantee	Laws define the responsibility division between the parties, project assessment, contract performance and change, conflicts in different paths, etc.
2. Approval guarantee	The government need to provide a legal guarantee to businesses involved in infrastructure projects
3. Police guarantee	The government needs to set up corresponding approval for security and incentives to encourage more businesses to participate in 3Ps infrastructure projects
4. Interest rate guarantee	The government should offer the recoverable rate of investment to businesses by offering a government-backed rate in the range of the given benefit level
5. Price guarantee	The government should offer the recoverable rate of investment to businesses by offering government-backed price in the range of the given benefit level
6. Purchase guarantee	The government should offer the recoverable rate of investment to businesses by offering government-backed purchase in the range of the given benefit level
7. Restriction of competition	Similar projects will not be built in a given time
8. Information transparency	Critical information about the 3Ps project disclosed by the government
9. Contract spirits	The government's contract spirits to guarantee the legitimate rights and interests of businesses
10. Contract spirits	The government ought to initially bear the risks of land policy changes, legal policy changes, and exchange rate changes

effectively needs to take into account distribution characteristics of the evaluative values. Suppose there is no difference in the values of a specific evaluative criterion for matching objects. In that case, the evaluative criterion has little effect on decision ranking, and thus, its weight should be given a smaller value accordingly. Otherwise, the evaluative criterion has a significant effect on decision ranking. Consequently, its weight should be given an immense value accordingly.

Based on the above analysis, a nonlinear optimization model is established to determine evaluative criterion weights through maximizing deviation between the values of the evaluative criteria for all matching objects as below:

$$\max\{d(B)\} = \max\sum_{q=1}^{n} d(B_q) = \sum_{q=1}^{n}\sum_{j=1}^{t}\sum_{k\neq j}^{t}\sqrt{\frac{1}{l}\sum_{\lambda=1}^{l}\left(\omega_j h_{qj}^{\delta(\lambda)} - \omega_k h_{qk}^{\delta(\lambda)}\right)^2}$$

$$(4.1)$$

$$s.t. \sum_{j=1}^{n}\omega_j = 1, 0 \leq \omega_j \leq 1, j = 1, 2, \ldots, t$$

Where $\omega_j, \omega_k (j, k = 1, 2, ..., t)$ denote evaluative criterion weights of the enterprise $B_q (q = 1, 2, ..., n)$ by the government $A_p (p = 1, 2, ..., m)$. Here, $d(B)$ denotes the sum of the deviation on Hesitant fuzzy element evaluative values of all the subjects in the enterprise by the government $A_p (p = 1, 2, ..., m)$. $h_{qj}^{\delta(\lambda)}$ and $h_{qk}^{\delta(\lambda)}$ are the λth elements in h_{qj} and h_{qk} with ascending sort order, respectively.

Calculating the compatibility on both sides. In the Multi-criteria Decision Making process, the government and enterprise satisfaction (i.e., compatibility) should be taken into full account and calculated to obtain an optimal matching scheme. In classic multi-attribute decision-making theory, evaluative values and evaluative criterion weights of schemes are weighted and summed to gain utility values. The schemes are ranked based on the utility values. Similarly, the enterprise's utility value in the enterprise is gained through weighted aggregating, the evaluative value, and the evaluative criterion weight of one subject in the enterprise. In general, the utility values and the compatibility of all the government subjects and the enterprise subjects have the same monotonicity. Based on the above analysis, the utility values and the compatibility of all the government subjects and the enterprise subjects can be calculated (Xia and Xu, 2011). Consequently, the compatibility matrices of all the government subjects and the subjects in the enterprise are obtained.

Establishing the matching DMs algorithm. After obtaining the compatibility matrices for all the government subjects and the enterprise's subjects, 0–1 decision-making variables $x_{ij} (i = 1, 2, ..., m; j = 1, 2, ..., n)$ are introduced to establish the matching decision model. If the value of the 0–1 decision variable is equal to 1, it denotes a matching between the government's subjects and the subjects in the enterprise. In contrast, if the value is equal to 0, it denotes non-matching between the subjects in the government and the subjects in the enterprise. During the Multi-criteria Decision Making process, both the government and the enterprise's subjects are supposed to hope to have greater satisfaction. Then, a Multi-criteria Decision Making model is established to maximize the sum of compatibility for all the subjects in the government and the subjects in the enterprise as follows:

$$\max Z_1 = \sum_{p=1}^{m} \sum_{q=1}^{n} \mu A_p, B_q x_{pq} \tag{4.2}$$

$$\max Z_2 = \sum_{p=1}^{m} \sum_{q=1}^{n} \mu' B_q, A_p x_{pq} \tag{4.3}$$

$$s.t. \sum_{q=1}^{n} x_{pq} = 1, \ x_{pq} \in \{0,1\}, p = 1, 2, ..., m \tag{4.4}$$

$$\sum_{p=1}^{n} x_{pq} \leq 1, \ x_{pq} \in \{0,1\}, q = 1, 2, ..., n \tag{4.5}$$

In the model above, the compatibility matrix of all the subjects in the government given by subjects in the enterprise and the compatibility matrix of all the subjects in the government's subjects are $u_{A,B}\left(\mu A_p, B_q\right)_{m \times n}$ and $u'_{B,A}\left(\mu' B_q, A_p\right)_{m \times n}$, respectively. Equations 4.2–4.3 are the objective functions of the model to maximize the compatibility of each pair between the subjects in the government and the subjects in the enterprise. Equation 4.4 is the constraint of the model. Since the number of subjects in the government does not exceed the number of subjects in the enterprise, Equation 4.4 is an equality constraint, which implies that one The question of government corresponds only to a possible question in government and individuals in compliance. At the same time, Equation 4.5 is an inequality constraint, which implies that one subject in the enterprise matches a potential subject in the government at most.

Solving the matching DMs algorithm. The matching decision model (Equations 4.2–4.5) is a multi-objective optimization problem that can be solved by many different algorithms, such as the firefly algorithms. This study used the method in Li and Wan to solve this problem. Let the above model's maximum values $Z_k^{\max}\,(k=1,2)$ be the objective function for considering a single-objective function separately. Then, the corresponding optimal solutions can be denoted as:

$$Z_k^{\max} = \max\left(\frac{1}{2}\left(\min\left\{\beta_k(x)|k=1,2\right\} + \frac{1}{2}\sum_{k=1}^{2}\beta_k(x)\right)\right)$$

(4.6)

$$s.t. \sum_{q=1}^{n} x_{pq} = 1, \sum_{p=1}^{m} x_{pq} \leq 1, x_{pq} \in \{0,1\}, p = 1,2,...,m; q = 1,2,...,n$$

Equation 4.6 is equivalent to a single-objective optimization as below:

$$Z_k^{\max} = \max\left(\frac{1}{2}\left(\min\left\{\beta_k(x)|k=1,2\right\} + \frac{1}{2}\sum_{k=1}^{2}\beta_k(x)\right)\right)$$

(4.7)

$$s.t.\,2\beta_k + \sum_{k=1}^{2}\beta_k = 4\beta, x_{pq} \in \{0,1\}, k = 1,2; p = 1,2,...,m; q = 1,2,...,n$$

Where β_k is the membership of the objective function $Z_k\,(k=1,2)$ (Wan and Li, 2013). Finally, using the linear weighted method [40] to calculate Equation 4.7, the optimal solutions of the matching optimization model (Equations 4.2–4.5) are obtained.

According to Equation 4.7, the results can be obtained as

$$Z_1^{\max} = 1.7521, Z_1^{\min} = 1.5589; Z_2^{\max} = 1.7877, Z_2^{\min} = 1.2945;$$

$$\beta_1 = \frac{Z_1 - 1.5589}{1.7521 - 1.5589}, \beta_2 = \frac{Z_2 - 1.2945}{1.7877 - 1.2945}.$$

Then, a single-objective linear programming model (SLPM) for the problem is presented as:

$$\max \beta$$

$$s.t.\, 2\beta_1 + \beta_1 + \beta_2 \geq 4\beta, 2\beta_2 + \beta_1 + \beta_2 \geq 4\beta, x_{pq} = 0$$

(4.8)

or

$$1;\, p = 1, 2, 3;\, q = 1, 2, ..., 6$$

It is easy to obtain the objective function values as $Z_1 = 1.6804, Z_2 = 2.0432$. An optimal solution is $X^* = \{(1,1), (2,2), (3,6)\}$. According to the gained optimal solution, the best matching scheme is $\{A_1 \leftrightarrow B_2, A_2 \leftrightarrow B_2, A_3 \leftrightarrow B_6\}$.

For a more detailed discussion, the following three different scenarios are considered.

Case 1: in this case, we examine the match-degree of *the government's subjects* in the three-dimensional space. For the subject A_1 in the government discussed, there are six possible matches, i.e., A_1 to B_1, B_2, B_3, B_4, B_5, and B_6, respectively. After A_1 matching is determined, A_2 can be matched with one of the five subjects left in the enterprise. Then A_3 can be matched with one of the four left. The corresponding results are shown in Figure 4.3. In this figure, different signs are utilized to show different matches. For example, the blue plots represent that subject A_1 matches subject B_1 in the enterprise; the red plots represent that A_1 matches subject B_2. The X-axis is the sum of the sequence numbers of A_2 and its matched subjects in the enterprise. For instance, taking one of the blue plots, the X-axis value is 6, which indicates that the subject B_4 in the enterprise matches with the subject A_2 as $4 + 2 = 6$.

Similarly, the Y-axis is the sum of sequence numbers of the subject A_3 and its matched subjects in the enterprise. Again, one of the blue plots is taken as an example. The Y-axis value is 9, indicating that the subject B_6 in the enterprise matches the subject A_3 as $6 + 3 = 9$. The Z-axis is the match-degree of subjects in the government. There are two different lines (i.e., blue line and red line) to present different government subjects' match-degree. As shown in Figure 4.3, the maximum match-degree of the subjects in the government is 1.7344; the corresponding matching scheme is $\{A_1 \leftrightarrow B_4, A_2 \leftrightarrow B_1, A_3 \leftrightarrow B_2\}$. The minimum match-degree of the subjects in the government is 1.5589; the corresponding matching scheme is $\{A_1 \leftrightarrow B_3, A_2 \leftrightarrow B_6, A_3 \leftrightarrow B_1\}$. The results show that the change of the match-degree of the subjects in the government is not significant as the matching scheme is changed.

Case 2: in this case, the enterprise subjects' match-degree in the three-dimensional space is discussed. For the subject A_1 in the government, there are six possible matches, i.e., A_1 to B_1, B_2, B_3, B_4, B_5, and B_6, respectively. After A_1 matching is determined, A_2 can be matched with one of the five subjects left in the enterprise. Then A_3 can be matched with one of the four left. The corresponding

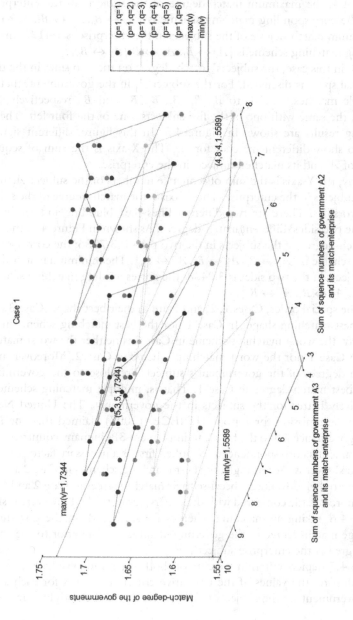

Figure 4.3 An algorithm based on the first case.

results are shown in Figure 4.4. In this figure, different signs are used to show different matches for A_1. The X-axis is the sum of the sequence numbers of A_2 and its matched subjects in the enterprise.

Similarly, the Y-axis is the sum of sequence numbers of the subject A_3 and its matched subjects in the enterprise. The Z-axis is the match-degree of subjects in the enterprise. There are two different lines (i.e., the blue line and the red line) to present the enterprise's subjects' different match-degrees. As shown in Figure 4.4, the maximum match-degree of the subjects in the enterprise is 2.0432; the corresponding matching scheme is $\{A_1 \leftrightarrow B_6, A_2 \leftrightarrow B_3, A_3 \leftrightarrow B_4\}$. The minimum match-degree of the subjects in the enterprise is 1.4132; the corresponding matching scheme is $\{A_1 \leftrightarrow B_1, A_2 \leftrightarrow B_3, A_3 \leftrightarrow B_6\}$.

Case 3: in this case, the subjects' match-degree on the two sides in the three-dimensional space is discussed. For the subject A_1 in the government, there are six possible matches, i.e., A_1 to B_1, B_2, B_3, B_4, B_5, and B_6, respectively. After matching, the same with one of the five subjects, one of the four left. The corresponding results are shown in Figure 4.5. In this figure, different signs are utilized to show different matches for A_1. The X-axis is the sum of sequence numbers of A_2 and its matched subjects in the enterprise.

Similarly, the Y-axis is the sum of sequence numbers of the subject A_3 and its matched subjects in the enterprise. The Z-axis is the match-degree of the subjects on the two sides. There are two different lines (i.e., blue line and red line) to present the two sides' different match-degrees. As shown in Figure 4.5, the maximum match-degree of the subjects in the two parts is 3.7236; the corresponding matching scheme is $\{A_1 \leftrightarrow B_1, A_2 \leftrightarrow B_3, A_3 \leftrightarrow B_6\}$. The minimum match-degree of the subjects in the two sides is 3.0438; the corresponding matching scheme is $\{A_1 \leftrightarrow B_6, A_2 \leftrightarrow B_3, A_3 \leftrightarrow B_4\}$.

With the special order, Cases 1, 2, and 3 are in their best shape. Case 3 is neither the best matching shape. In Case 1 nor the best matching scheme in Case 2. Similarly, the worst matching scheme in Case 3 is neither the worst matching scheme in Case 1 nor the worst matching scheme in Case 2. Moreover, multiple match-degrees of the government's subjects are close to the government's subjects' best match-degree in Case 1. That is, multiple matching schemes are comparatively better for the subjects in the government. The United Nations Economic Commission for Europe (UNECE) (2007) claimed that the inefficiency in governance led to the failure to implement 3Ps in many countries. Thus, the government's correspondence was ranked first as a necessary factor to ensure 3Ps projects' success. Accordingly, the best matching schemes in Case 2 and Case 3 fully play to their advantages because their match-degrees in Case 2 and Case 3 become more volatile compared with the others, respectively. However, as shown in Figure 4.6, taking all matching schemes for subject A_1 in the government, the average match-degree of the government subjects is superior to the average match-degree of the enterprise subjects.

Figure 4.7 depicts criterion weights on both parties involved. Precisely, the left side displays the values of the evaluative criterion weights for each subject in the government by the subjects in the enterprise; and the right side displays

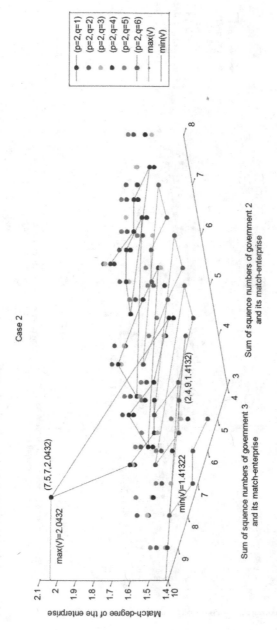

Figure 4.4 An algorithm based on the second case.

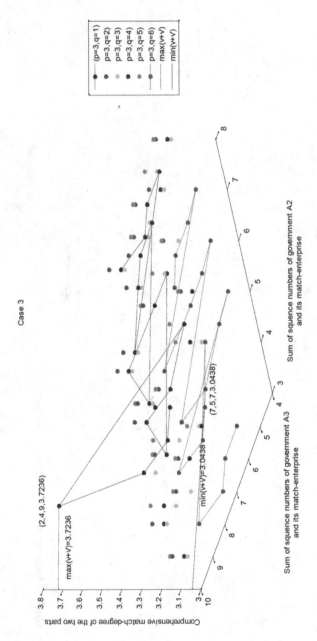

Figure 4.5 An algorithm based on the third case.

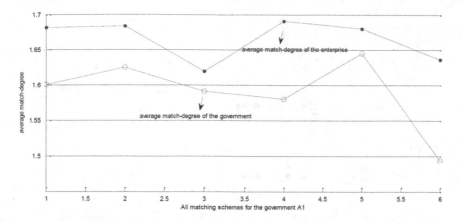

Figure 4.6 Mean of weights.

the values of the evaluative criterion weights for each subject in the enterprise by the subjects in the government. The results showed that the evaluative criterion weights of the three subjects in the government are very near and lie at 0.2000. In comparison, the evaluative criterion weights of the six subjects in the enterprise remained stable, despite a large increase in the third evaluative criterion of the subject B_6 in the enterprise and a large decrease in the first evaluative criterion of the subject B_2 in the enterprise. The average value of evaluative criterion weights on the right side is higher than the average value of evaluative criterion weights on the left side. It reveals that government intervention and corruption may be the significant obstacles to 3Ps projects' success in China.

4.6 Evaluating 3Ps

As indicated above, 3Ps performance evaluation/measurement is one of the popular topics in current 3Ps research. Theoretically, performance evaluation is a process that is designed for examining and reporting the effectiveness and efficiency of the actions taken towards defined objectives in organizations (Neely et al., 2005). Solomon and Young (2007) maintain that strategic objectives act as foundations of evaluation. In 3Ps, realizing value for money (V*f*M) has been acknowledged as the principal benchmark of the projects' strategic objective (Akintoye et al., 2003; Grimsey and Lewis, 2005; Liang et al., 2018; Zhang et al., 2019). The Treasury Taskforce (1998) supports this perspective and argues that 3Ps should provide better value for money than traditional procurements. The value for money is a subjective but critical concept, and it can be defined as 'the optimum combination of whole-life cost and quality to meet the user's requirement' (Office of Government Commerce, 2002, p. 6).

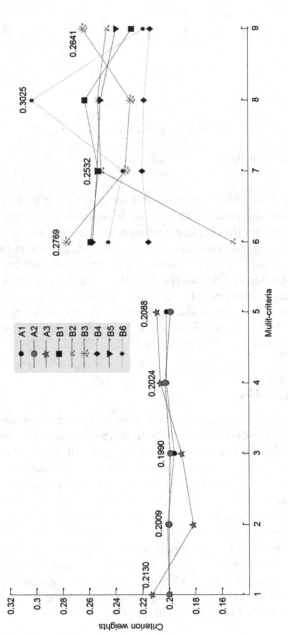

Figure 4.7 Conversion of weights based on indicator.

The value for money of 3Ps is determined by the *Public Sector Comparator* (PSC), especially in Australia and the UK (Treasury Taskforce, 1999). The public section comparator compares the proposed cost using 3Ps and the benchmark cost of the project that is estimated based on the application of traditional procurement (Grimsey and Lewis, 2005). It is worth noting that the public section comparator is concerned with a cost; therefore, the evaluations of 3Ps in Australia and the UK have been oversimplified (e.g., National Audit Office – NAO, 2000; Fitzgerald, 2004; Allen Consulting Group – ACG, 2007). In the rest of the world, such as the United States, South America, Africa, New Zealand, Asia, and other EU countries, cost performance dominates in 3Ps evaluations as well (e.g., Haskins et al., 2002; Sachs et al., 2005; Anastasopoulos et al., 2010). However, 3Ps' values cannot be entirely revealed by financial issues but by project completion time (Yong, 2010). Thus, several studies examine 3Ps' performance by focusing on meeting the projects' schedules (e.g., NAO, 2003; Zietlow, 2005; Raisbeck et al., 2010).

Evaluating 3Ps under the objective of V*f*M is complex, and the use of absolute 'cost' and 'time' measurements cannot reflect the complexity associated with 3Ps delivery (Office of Government Commerce, 2002; European Commission, 2003). Amos (2004) suggests that technical and allocative efficiency, in addition to 'cost' and 'schedule' measures, should be involved in the project evaluation of a 3Ps. Table 4.4 provides a summary of the critical studies on the performance evaluations of 3Ps.

The measures above (e.g., cost, time, technical efficiency, and allocative efficiency), however, are still insufficient to fully capture value for money, because such critical but qualitative issues as crucial project stakeholders' satisfaction (e.g., public client, shareholders, and end-users) cannot be examined properly (Henjewele et al., 2011; Liu et al., 2014b). For that matter, Yuan et al. (2009) develop an innovative Key Performance Indicators System (KPIS) for 3Ps, and this system consists of *five* key measures that are comprised of a sequence of key performance indicators, involving: (1) *project physical characteristics*, (2) *financing*

Table 4.4 Critical studies on the performance of 3Ps

Measures for evaluations	Key relevant studies	Nature of evaluation
Cost (budget)	NAO (2000), Haskins et al. (2002), Pakkala (2002), Fitzgerald (2004), Liautaud (2004), Zietlow (2005), Sachs et al. (2005), ACG (2007), and Anastasopoulos et al. (2010)	*Ex-post evaluation*
Cost and time (schedule)	Mott MacDonald (2002), NAO (2003), and Raisbeck et al. (2010)	*Ex-post evaluation*
Technical and allocative efficiency	Amos (2004)	*Ex-post evaluation*

and marketing indicators, (3) *innovation and learning indicators,* (4) *stakeholders' indicators,* (5) *process indicators.*

Despite the innovation of the key performance indicators of Yuan et al. (2009), some essential aspects of the theoretical framework of performance measurement have been largely neglected. Take stakeholder indicators, for example. Neely et al. (2001) argue that the relationship between an organization and its stakeholders is reciprocal, and performance measurement should examine stakeholders' satisfaction and contribution to the organization. Moreover, 3Ps' strategic objective (i.e., value for money) has not been addressed in the current approaches/ frameworks used for evaluating the projects. However, it is acknowledged as the baseline of performance evaluations (Liu et al., 2014b).

The current research on 3Ps evaluation, as mentioned above, is primarily concerned with an *ex-post* evaluation that is conducted after the completion of construction, while few studies focus on developing approaches, mechanisms, or frameworks for evaluating the interception stages (i.e., pre-design and pre-construction stages) of a 3Ps, which is usually referred to as *ex-ante* evaluation. With this in mind, this chapter will bridge the knowledge gaps, mentioned earlier, in 3Ps evaluations.

Public–Private Partnerships are innovative methods used by the public sector to contract with the private sector, which brings its capital and ability to deliver projects on time and on budget. In contrast, the public party retains the accountability to provide these services to the public to advantage the public and deliver economic development and improved quality of life (Baizakov, 2008). 3Ps projects' worldwide popularity is justified because 3Ps can effectively avoid the often-negative effects of either exclusive public ownership and delivery services or outright privatization on the other hand. Furthermore, 3Ps combine both entities' best: the public sector with its regulatory actions and protection of the public interest; and the private sector with its resources, management skills, and technology. 3Ps' main goals include financing, designing, implementing, and operating public-sector facilities and services. The key characteristics associated with 3Ps encompass

1) Long-term (typically up to 30 years) service provisions.
2) The transfer of risk to the private sector.
3) Different forms of long-term contracts are drawn up between legal entities and public authorities.

Although the types of 3Ps vary due to governments' different needs for infrastructure services, two broad categories of 3Ps have been identified: the institutionalized kind that refers to all joint ventures among public and private stakeholders; and contractual 3Ps. More specifically, 3Ps can take many forms and may incorporate part or all of the following features (Peirson and McBride, 1996; Thobani, 1998):

1) The public sector transfers facilities controlled by itself to the private sector (with or without payment in return) generally for the term of the arrangement.

2) The private sector builds, extends, or renovates a facility.
3) The public sector specifies the operating features of the facility.
4) Services are provided via the private sector using the facility for a defined period (usually with restrictions on operations and pricing).
5) The private sector agrees to transfer the facility to the public sector (with or without payment) at the end of the arrangement.

In the presence of hundreds of 3Ps projects' applications, fundamental decision of the "managers" to analyze them and select suitable projects for implementation. Motivated by the observation that the evaluation of 3Ps projects consists of multiple criteria, this section proposes a decision approach that minimizes the total deviation from the ideal point to determine the weights associated with each criterion.

4.7 Developing a method for 3Ps

4.7.1 Choosing an Algorithm

Infrastructure 3Ps are construction projects in nature. The construction sector is a project-led industry, where companies have to meet the goals at both organization and project levels (Love and Holt, 2000). Hence, performance measurements in construction typically focus on three levels: (1) industry; (2) corporate; and (3) project (Elyamany et al., 2007), with the emphasis being placed on KPIs and performance measurement systems (PMSs) (Bassioni et al., 2004). Theoretically, 'key performance indicators are measures that are indicative of the performance of an associated process' (Beatham et al., 2004, p. 106), while 'a performance measurement systems is a structure in which strategic, tactical and operational actions are linked to process to provide the information required to improve the program or service on a systematic basis' (del-Rey-Chamorro et al., 2003, p. 47).

One principle under the performance measurement is that the developed performance measures must reflect the context in which they are applied (Neely, 1999). Infrastructure 3Ps projects are undertaken by *Special Purpose Vehicles* (SPVs), which are the consortiums responsible for designing, building, financing, operating, and maintaining public assets over a concession period (Zheng et al., 2008). Special purpose vehicles exhibit a dual character. This implies that they operate in a context where goals and objectives at organization and project levels must be achieved; therefore, the framework used for evaluating 3Ps should cover the issues at both corporate and project levels (Liu et al., 2014b).

Although key performance indicators are popular in the construction industry, they examine the project rather than company factors. Corporate-related issues (i.e., strategies) cannot be adequately addressed in this framework (Kagioglou et al., 2001). Moreover, due to the lagging nature, key performance indicators provide limited assistance for decision-making (Bassioni et al., 2004), which is highly important for *ex-ante* project evaluation. The performance measurement systems integrated with the project perspective are deemed to be more appropriate for 3Ps than sole key performance indicators. They are robust enough to capture project and corporate issues and provide insight into investment decision-making (Kagioglou et al., 2001).

4.7.2 Balanced Scorecards and 3Ps

During the last two decades, several performance measurement systemshave been widely applied in construction. The quality-based excellence models and Balanced Scorecard (BSC) are undoubtedly the most reputable systems (Robinson et al., 2005). Nevertheless, many criticisms of them were raised by researchers. On the one hand, the quality-based excellence models are vague and rarely focus on such vital areas as innovation, improvement, supplier partnership, and strategic positioning (Leonard and McAdam, 2002). On the other hand, the balanced scorecard is too narrow to examine the factors that are critical to the performance and success of the organization, as it involves only customers and shareholders but ignores a wide range of essential stakeholders, e.g., suppliers, alliance partners, employees, regulators, and local community (Neely et al., 2001). Further, the balanced scorecards fails in highlighting 'the relationship between the measures proposed for certain goals' (Kagioglou et al., 2001, p. 87).

According to Leiringer (2006), innovation (e.g., technical, management, and evaluation) and strategies play a decisive role in 3Ps. Additionally, Special purpose vehicles incorporate a set of entities, including public authority, concessionaire, and constituent members (Chinyio and Gameson, 2009); thus, effective integration of multiple stakeholders is essential for successfully delivering a 3Ps project (Kwak et al., 2009; Yong, 2010). However, the excellence models and balanced scorecards have inherent deficiencies in capturing innovation and strategies and examining organizations in a multiple-stakeholder environment, respectively. Hence, both of them might not be suitable choices for evaluating infrastructure 3Ps.

Based on the characteristics of 3Ps, a framework that is capable of evaluating outputs at the project and corporate levels simultaneously to achieve the long-term strategic objective and deal with the complexities raised by multiple stakeholders is an ideal tool to underpin the performance measurement of 3Ps projects (Liu et al., 2014b). With this in mind, the Performance Prism proposed by Neely et al. (2001) is deemed a feasible one. This is because the Performance Prism is a holistic framework structured to shed light on the complexity of multiple-stakeholder integration and provide assistance in directing and guiding performance measurement for long-term success within a particular business environment (Neely et al., 2002). Its robust ability in measuring organizational performance, especially in multiple-stakeholder environments, has been validated by a case study for the logistics industry. The Performance Prism is comprised of *five* interrelated facets of measurement (Neely et al., 2001, pp. 6–7) (see Figure 4.1):

A. Stakeholder satisfaction: who are our stakeholders, and what do they want and require?
B. Strategies: what strategies do we require to satisfy these sets of wants and needs?
C. Processes: what processes do we need to allow our strategies to be delivered?
D. Capabilities: what capabilities (people, practice, technology, and infrastructure) do we need to implement to operate our processes?
E. Stakeholder contribution: what do we want and require from our stakeholders?

The views of stakeholders incorporated in the Performance Prism mean to give the ability to overcome the hurdle triggered by multiple stakeholders in 3Ps evaluation. The facet of stakeholder satisfaction in this framework is broader than other performance measurement systems (such as the balanced scorecards). It evaluates shareholders and customers and suppliers, alliance partners, and even intermediaries (Neely et al., 2001). At the project level, 'stakeholders are individuals or organizations that are affected by or affect the development of the project' (El-Gohary et al., 2006, p. 595). Various parties other than the public client and leading concessionaire are involved in 3Ps, such as sponsors, subcontractors, suppliers, and facility management organizations (Chinyio and Gameson, 2009). Considering these perspectives, the *Stakeholder Satisfaction* of the Performance Prism can completely cover the stakeholders who will be critical to a 3Ps's success. In other words, this facet is valuable for supporting 3Ps agencies to identify what key stakeholders will be involved in the projects and how to fulfill and meet their needs and expectations, both of which are crucial for a 3Ps *ex-ante* evaluation.

This framework also indicates the importance of stakeholder contribution in evaluation, which is neglected by other performance measurement systems (Neely et al., 2002). Essentially, 'not only do organizations have to deliver value to their stakeholders but also organizations enter into a connection via their stakeholders which should involve the stakeholder contributing to the organization' (Neely et al., 2001, p. 7). This implies that the organization's performance is substantially affected by stakeholders' satisfaction and contribution to the organization. In 3Ps, Special purpose vehicles consist of a sequence of parties, resulting in the project managers needing to oversee a wide range of group performance to maintain satisfactory progress. Therefore, the performance of a 3Ps is determined by how the key stakeholders satisfy the organization and how adequately they contribute to the organization or the project. Thus, the *Stakeholder Contribution* is important for 3Ps, and it can help identify what areas the key stakeholders should provide support in the projects' *ex-ante* evaluation.

Apart from stakeholder-related issues, the Performance Prism emphasizes the necessity of 'strategies' and 'capabilities.' Strategy, as discussed above, underpins an organization's operations. Neely et al. (2001) claim that the *Strategies* perspective in the Performance Prism indicates the short-run and long-term strategies that the organization should implement. So, this facet fits nicely into the early-stage evaluation of 3Ps. It enables the supporting agencies to clearly understand the direction and strategic objective (i.e., the project's achievement of Value for Money).

Conversely, the balanced scorecards fails to address this critical issue, 'seeming to be assuming that all measures will be specific only to a particular goal' (Kagioglou et al., 2001, p. 87). This deficiency can also be found in Yuan et al.'s (2009) proposed KPIS. Furthermore, any business operation in an organization must be supported by specific skills, practical procedures, physical infrastructure, and technologies known as organizational capabilities (Neely et al., 2002). In practice, a 3Ps project's delivery depends upon Special purpose vehicles' capabilities, including a skilled workforce, finance infrastructure, technologies, and equipment (Yong, 2010). Hence, the *Capabilities* facet can provide public authorities with an insight into the resources and skills required for the asset's procurement and assist them in identifying what kind of private entity is suitable for the proposed projects.

In the remaining facet of the Performance Prism, the *Process* can also operationalize with infrastructure 3Ps effectively. As proffered by Neely et al. (2001), the *Process* facet is useful for examining the *Process*'s effectiveness and efficiency to send the businesses towards its strategies. The *Process* of a 3Ps, normally, is dynamic and complex, and it is critical to the successful delivery of the project (EIB, 2012; Liu et al., 2014d). Therefore, the *Process* perspective is vital for evaluating 3Ps' interception stages, and it helps launch an effective and efficient process for the projects.

4.7.3 Key Concepts of Performance in 3Ps

The previous discussion suggests that the Performance Prism is feasible for 3Ps *ex-ante* evaluation. Figure 4.8 illustrates how the prism framework can match the characteristics of 3Ps projects. A series of core indicators (CIs) will be proposed within the Performance Prism context to enhance the applicability.

The early stages of an infrastructure 3Ps can be divided into three phases, *project identification*, *detailed preparation*, and *procurement*, in which a wide range of work must be completed in *ex-ante* evaluation such as project identification, output definition and specification, assessment for value for money, feasibility study, analysis for affordability and bankability, risk evaluation (e.g., identification, assessment, and allocation), and some further studies (i.e., bid evaluation criteria) (EIB, 2004). Table 4.5 summarizes the core indicators for evaluating 3Ps' early stages based on these points of view.

The CIs presented above are derived from the normative literature of 3Ps (Yuan et al., 2009; EIB, 2004; Garvin et al., 2011; Liu et al., 2014a, b), and all of them are associated with the critical works of *ex-ante* evaluation in 3Ps (e.g., studies on feasibility, affordability and bankability, risk analysis, macro-environment assessment, and identification of required resources and key stakeholders' expectation and contribution). Thus, these derived CIs under the Performance Prism can serve as an effective and efficient mechanism and performance measurement

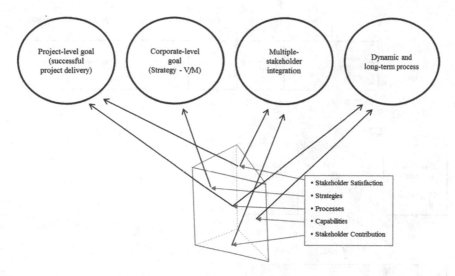

Figure 4.8 3Ps and the performance cylinder (Adapted from Neely et al., 2001).

Table 4.5 The Performance cylinder core indicators

Measurement facets	Core indicators (CIs)
Stakeholder Satisfaction (F1)	Public client's satisfaction (CI_{F1-1})
	Main concessionaire's satisfaction (CI_{F1-2})
	Creditors' satisfaction (CI_{F1-3})
	Shareholders' satisfaction (CI_{F1-4})
	Subcontractors' satisfaction (CI_{F1-5})
	End-users' satisfaction (CI_{F1-6})
Strategies (F2)	Value for money (CI_{F2-1})
Process (F3)	The comprehensiveness of environment analysis (CI_{F3-1}) (e.g., political, macroeconomic, social, and legal)
	Appropriateness of service and output definitions (CI_{F3-2})
	The comprehensiveness of feasibility study (CI_{F3-3})
	Effectiveness of the analysis for affordability and bankability (CI_{F3-4})
	Effectiveness of risk evaluation (CI_{F3-5})
	Appropriateness of bid evaluation criteria (CI_{F3-6})
Capabilities (F4)	Skilled project and advisor teams (CI_{F4-1})
	Finance infrastructure of potential concessionaire (CI_{F4-2})
	Innovation ability of potential concessionaire (CI_{F4-3})
Stakeholder Contribution (F5)	Governance and support from a public authority (CI_{F5-1})
	Concessionaire's contribution to asset sustainability (CI_{F5-2})
	End-users' willingness to use the asset to be procured (CI_{F5-3})

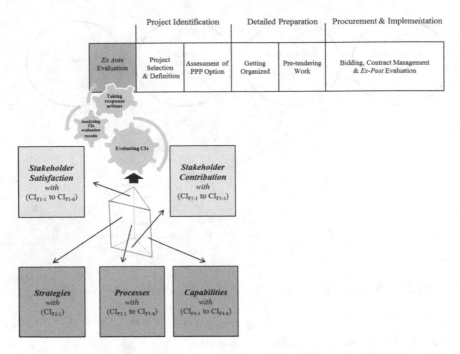

Figure 4.9 Multiple criteria 3Ps (Adapted from Liu et al., 2014b).

systems that can ensure public authorities/support agencies perform well in evaluating the interception stages of infrastructure 3Ps projects (see Figure 4.9).

4.8 A solution for evaluating 3Ps

Effective evaluation of the financial value of 3Ps contracts is a multifaceted task due to the long improvement horizon of 3Ps contracts, risk related to the evolutions of prices, the impact on state revenue, loss of natural resources, natural disorder, cultural effect, and uncertainty of the operating solutions and legal evolutions. The main strategy for analyzing 3Ps contracts by public managers is the use of benefit/cost ratios (B/C), discounted cash flow (DCF) strategies such as the net present value (NPV), and internal rate of return (IRR). The main deficiency with these strategies is the failure to evaluate the managerial flexibility induced by uncertainty.

Substantial amounts of natural resources and public goods with great value are presently handled in true concession or Public–Private Partnerships. Governments in Eastern Europe are frequently criticized for not measuring their total value accurately and unnecessarily wasting this resource. It is often repeated in the context that governments cannot convincingly demonstrate sufficient return on investment or value of such resources they own, which is mainly because it does not rely on robust measurement strategies. This is largely due to how we see some fundamental concepts of cash flow, profitability, and profit at the government level. Lack of notions of benefit and profitability or their understanding are visible in regulatory acts governing the rules in the decisions on concessions, Public–Private Partnerships, and public investments. As companies use higher investment, the state remains locked in a set of static, outdated, and insufficient to fix decisions that govern economic results over an extended period.

Our technique starts by examining the financial issue that raises the scores following a dynamic uncertainty. In our opinion, the lack of information about the future makes any initial negotiation that concludes with a fixed price a mistake in negotiating any lease. Looking at concessions from two perspectives connected to uncertainty and time irreversibility of the contract, we propose a specific condition to evaluate it and, finally, its price using real options' methodological strategy. A real examination of the financial attractiveness of 3Ps contracts is a very difficult task, especially because of some unique sources like the long improvement time of the 3Ps contracts, the uncertainty of future prices and costs, impact on state revenue, loss of natural sources, social impact, and risk of the future legal operating conditions. The main strategy for the evaluation of 3Ps contract decisions by public managers is the use of benefit/cost ratios, traditional discounted cash flow techniques like the net present value, land expectation value (LEV in the case of forestry exploitations), and internal rate of return. The main deficiency of these techniques is the failure to evaluate the managerial flexibility induced by uncertainty.

In this chapter, we present an instance of using the real options approach to model flexibility and uncertainty in gold mining investment analysis from the perspective of the value created for the state via a concession contract.

As companies use total higher investment, the state remains locked in a set of static, outdated, and insufficient to fix the decisions that govern financial results over a very extended period. These things are made worse by a lack of information on the development of certain economic indicators associated with developments of different raw material prices due to a focus only on macroeconomic indicators and a lack of monitoring indicators allowing the state to become a decision-maker that relies on the same means they utilize and private companies.

Given that the private economic environment is constantly developing new concepts and the companies are constantly developing their decision-making systems, it is up to the government as a decision-making body to adapt its organizational practices to compete and negotiate on the same foundation with competing companies or partners.

Our strategy starts from examining the financial issue that raises scores following a dynamic uncertainty. In our perspective, the lack of information about the future makes any initial negotiation that concludes with a fixed price a mistake in negotiating any lease. In other words, the concession decision should be viewed as an investment with a high degree of irreversibility. Looking at concessions from both perspectives related to the contract's uncertainty and time irreversibility, we propose a specific solution to evaluate the specific solution and, finally, its price utilizing real options' methodological technique.

This strategy allows us to capture the connection among the government or any other public resources manager and the private concessionaire in terms of concession value created for both the state and private concessionaire.

Analyzing the state's natural resources, we should first assess two alternative actions that the government has at its disposal: the direct use of natural resources and leasing them to private exploitation. The appeal to the concession system must be justified only by how leasing development brings more benefits than direct investment. This is possible only if the private management efficiency is higher than that resulting from direct exploitation by the state.

Usually, concession contracts and Public–Private Partnerships are, in the best-case scenario, evaluated based on two principles: a cost-benefit analysis based on the net present value criteria and political considerations. The uncertainty valuations represent, at best, only a heuristic approach based on risk factor identification but with no quantitative model to analyze the impact of risk on project value. This approach is insufficient in understanding the impact of the risks on project value. It does not offer a correct analysis of the dynamic decision-making possibilities offered by the flexibility induced by uncertainty and the potential for further learning. Suppose the public sector is unaware of the investment project's options values related to a concession contract. In that case, the public sector's significant options value may be lost in favor of private companies.

4.9 A Theoretical Strategy

Effective evaluation of the financial value of 3Ps contracts is a very complex task due to, among other reasons, the long development horizon of 3Ps contracts, uncertainty of future prices and costs, effect on state revenue, loss of natural resources, social impact, and uncertainty of the future legal operating conditions. The main approach to evaluating concession contract decisions by public managers is the use of benefit/cost ratio, traditional discounted cash flow techniques such as the net present value, and internal rate of return. The main deficiency with these techniques is the failure to evaluate the managerial flexibility induced by uncertainty.

This investment evaluation method's main problem is that all this analysis is linear and static and assumes that either the investment opportunity is reversible or irreversible. It is a now-or-never opportunity (Dixit and Pindyck, 1994). These limitations render the conclusions of DCF valuation somewhat suspect, and using it can under-analyze the concession value.

Private managers tend to use increasingly new investment appraisal methods. If the state remains limited to static methods, the concession contracts' loss can be significant in the long term. Evaluating specific literature, we can find that concession and Public–Private Partnerships are extensively debated from the point of view of private management using real options theory. In other words, few research papers evaluate the concession and 3Ps investment projects as real options from the state perspective. Even more national regulations and European regulations do not reference real options valuation when referring to evaluating this kind of contract and the value brought for the state.

New and more convincing valuation of investments and decision-making in projects can be done according to real options theory, as argued by Dixit and Pindyck (1994); Trigeorgis (1996); and Copeland and Antikarov (2001). In our opinion, this kind of valuation should not be avoided by the state when it tries to evaluate its revenue from concession contracts.

Classical real options models try to recognize a linked portfolio of trading assets with the same risk as the project and the same returns (cash flows). In this case, the project cash flows are evaluated by identifying a portfolio trading strategy that generates the same cash flows as the project's cash flow.

Because it is very difficult to identify a portfolio of traded assets that exactly imitate risks and returns of the investment project, and sometimes it is impossible to do so, Copeland and Antikarov (2001) proposed a simpler strategy free from any market-traded asset. This new simpler approach is associated with the marketed asset disclaimer (MAD) assumption. The marketed asset disclaimer technique uses the current value of a project without options as the best unbiased estimate for the market value of the project (Copeland and Antikarov, 2001). The marketed asset disclaimer assumption eliminates the difficulty of recognizing a twin portfolio of linearly independent securities that can generate the same risk and cash flow as the project.

The analysis uses standard tools of option pricing and investment in discrete time, using the multiplicative binomial technique (Cox et al., 1979) based on the marketed asset disclaimer assumption. The binomial process was chosen because it is the simplest of the option pricing formulas (Elton and Gruber, 1995). In the binomial model, we can check the likelihood of developing exercise at every point of the option's lifetime. Besides, the binomial model does not depend on the prospect of certain special conditions. This implies that the real options model is not dependent on investors with subjective probabilities about an upward and downward movement in the underlying asset. These probabilities are not the same as those of the project.

The model requires the introduction of limitations that will allow the state to control the outcome if the beneficiary's financial results are better than the concession's initial expected value.

VL is the maximum financial assessment imposed by the concession or 3Ps. Concerning the type of restriction, imposed VL can be determined in the following way:

Let L be the upper limit of profitability imposed by the concession contract. In these conditions, VL can be determined in the following way:

$$V_L = \begin{cases} L + \left(Vu^i d^j - L\right) * p_a, & L \prec Vu^i d^j \\ Vu^i d^j, & L \succ Vu^i d^j \end{cases} \tag{4.9}$$

Where p_a is the share retained by the private sector concerning the excess yield results and V_c is the amount transferred to the state project due to exercising the option of limiting the yield.

In these solutions, assessing the value of the concession made by the state becomes dependent on the investment objective's operation. It can be determined using the same reasoning used by private companies when turning to real options theory.

In this context, the value resulting from the concession or 3Ps will be dependent on two sets of options that are accessed at the same time by the state and by the private sector. Under these conditions, the value of the project is connected to both decision-makers. This fact compels their cooperation in the future to ensure optimum value for each of the partners of the concession contract.

4.10 Proposed approaches

The multiple-criteria 3Ps projects evaluation and selection problem is formulated as follows:

$A = (A_1, A_2, ..., A_m)$: a set of m projects.

$C = (C_1, C_2, ..., C_n)$: a set of n criteria.

$\Upsilon = \left[y_{ij} \right]_{mn}$: the decision matrix where y_{ij} is the input data for the project i concerning criterion j, $i = 1, 2, ..., m$; $j = 1, 2, ..., n$.

The decision matrix $\Upsilon = \left[y_{ij} \right]_{mn}$ is normalized to the matrix $X = \left[x_{ij} \right]_{mn}$ using the following formula:

$$x_{ij} = \frac{y_{ij} - y_j^{min}}{y_j^{max} - y_j^{min}}, \text{ for benefit criteria.}$$

$$x_{ij} = \frac{y_j^{max} - y_{ij}}{y_j^{max} - y_j^{min}}, \text{ for cost criteria.}$$

Where $y_j^{max} = \max\{y_{1j}, y_{2j}, ..., y_{mj}\}$, and $y_j^{min} = \min\{y_{1j}, y_{2j}, ..., y_{mj}\}$.

$W = (w_1, w_2, ..., w_n)$: a set of criteria weights, and $\sum_{j=1}^{n} w_j = 1$.

Min, the total deviation from the ideal point, is intuitively appealing since the principle behind this approach is seeking to make all derived scores as close to the ideal point as possible. In line with Ma et al. (1999, 2020), the matrix $X = \left[x_{ij} \right]_{mn}$ is transformed into a weighted decision matrix $\Psi = \left[z_{ij} \right]_{mn}$, where

$$z_{ij} = x_{ij} w_j, i = 1, 2, ..., m, j = 1, 2, ..., n. \tag{4.10}$$

The ideal point is defined as $\Omega^* = \left\{ z_1^*, z_2^*, ..., z_n^* \right\}$, where

$$
\begin{aligned}
z_j^* &= \max \left\{ z_{1j}, z_{2j}, ..., z_{mj} \right\} \\
&= \max \left\{ x_{1j} w_j, x_{2j} w_j, ..., x_{mj} w_j \right\} \\
&= \max \left\{ x_{1j}, x_{2j}, ..., x_{mj} \right\} w_j \\
&= x_j^* w_j,
\end{aligned} \tag{4.11}
$$

and x_j^* is the ideal value under criterion j. It is straightforward because the maximal scores obtained from certain criteria are reasonably regarded as the target to reach (Grimsey and Lewis, 2002). The following function can measure the discrepancy between the performance under certain criteria and the ideal point:

$$
\begin{aligned}
D_i &= \sum_{j=1}^{n} (z_{ij} - z_j^*)^2 \\
&= \sum_{j=1}^{n} (x_{ij} - x_j^*)^2 w_j^2.
\end{aligned} \tag{4.12}
$$

A multi-objective programming model is presented below to optimize the performance of all projects:

$$
\begin{cases}
f_1 = \min \sum_{j=1}^{n} (x_{1j} - x_j^*)^2 w_j^2 \\
f_2 = \min \sum_{j=1}^{n} (x_{2j} - x_j^*)^2 w_j^2 \\
\quad \cdots \\
f_m = \min \sum_{j=1}^{n} (x_{mj} - x_j^*)^2 w_j^2 \\
s.t. \sum_{j=1}^{n} w_j = 1,
\end{cases} \tag{4.13}
$$

which is converted into a single-objective optimization model:

$$
\begin{cases}
\min F = \sum_{i=1}^{m} \sum_{j=1}^{n} (x_{ij} - x_j^*)^2 w_j^2 \\
s.t. \sum_{j=1}^{n} w_j = 1.
\end{cases} \tag{4.14}
$$

To solve the quadratic programming in Equation 4.13, we construct a Lagrange function using a Lagrange multiplier λ:

$$L = \sum_{i=1}^{m}\sum_{j=1}^{n}\left(x_{ij}-x_{j}^{*}\right)^{2}w_{j}^{2}+\lambda\left(\sum_{j=1}^{n}w_{j}-1\right). \tag{4.15}$$

The Hessian matrix of Equation 4.15 concerning w_j is an $n \times n$ diagonal matrix, and its diagonal elements are $2\sum(x_{ij}-x_{j}^{*})^{2}\geq 0$. Therefore, the Lagrange function has a minimum value, which is derived by differentiating Equation 4.15 for w_j and λ respectively:

$$\begin{cases} \lambda^{*} = \dfrac{1}{2\displaystyle\sum_{j=1}^{n}\left[\displaystyle\sum_{i=1}^{m}(x_{ij}-x_{j}^{*})^{2}\right]^{-1}}, \\[4mm] w_{j}^{*} = \dfrac{1}{\displaystyle\sum_{j=1}^{n}\left[\displaystyle\sum_{i=1}^{m}(x_{ij}-x_{j}^{*})^{2}\right]^{-1}\displaystyle\sum_{i=1}^{m}(x_{ij}-x_{j}^{*})^{2}}. \end{cases} \tag{4.16}$$

Since the constraint of Equation 4.14 is a non-empty convex set, and the objective function is convex, the optimal solutions to Equations 4.16 to 4.14 are the optimal global solutions.

4.11 Numerical illustration

To illustrate the effectiveness of the proposed approach, we consider the situation that 3Ps projects involve multiple criteria, namely, indirect economic contribution (y_{i1}), direct economic contribution (y_{i2}), technical contribution (y_{i3}), social contribution (y_{i4}), and scientific contribution (y_{i5}). A summary of the input data is presented in Table 4.6.

Using Equation 4.16, the weights associated with different criteria are:

$$W = (0.1786, 0.2487, 0.2222, 0.1521, 0.1985). \tag{4.17}$$

Therefore, each project's scores and rankings are reported in the seventh and eighth columns of Table 4.6 and Figure 4.10.

It is observed that our approach, namely, minimizing the total deviation from the ideal point, considers the direct economic contribution as the most important criterion and the social contribution as the least important one. Furthermore, our approach provides a complete ranking of all projects, reflecting the proposed approach's discriminatory power.

3Ps refers to the public sector's innovative methods to contract with the private sector, which brings its capital and ability to deliver projects on time to

Table 4.6 Multiple criteria 3Ps

Projects	Criteria					Scores	Ranking
	y_{i1}	y_{i2}	y_{i3}	y_{i4}	y_{i5}		
P1	67.53	70.82	62.64	44.91	46.28	0.8906	2
P2	58.94	62.86	57.47	42.84	45.64	0.8040	3
P3	22.27	9.68	6.73	10.99	5.92	0.0016	15
P4	47.32	47.05	21.75	20.82	19.64	0.3684	12
P5	48.96	48.48	34.9	32.73	26.21	0.5080	11
P6	58.88	77.16	35.42	29.11	26.08	0.6237	8
P7	50.1	58.2	36.12	32.46	18.9	0.5167	10
P8	47.46	49.54	46.89	24.54	36.35	0.5645	9
P9	55.26	61.09	38.93	47.71	29.47	0.6625	5
P10	52.4	55.09	53.45	19.52	46.57	0.6489	6
P11	55.13	55.54	55.13	23.36	46.31	0.6806	4
P12	32.09	34.04	33.57	10.6	29.36	0.3220	14
P13	27.49	39	34.51	21.25	25.74	0.3555	13
P14	77.17	83.35	60.01	41.37	51.91	0.9636	1
P15	72	68.32	25.84	36.64	25.84	0.6283	7

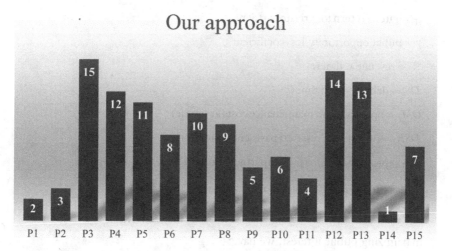

Figure 4.10 Ranking of the 3Ps.

budget. In contrast, the public sector retains the responsibility to provide these services to the public in a way that benefits the public. It delivers economic development and improved life quality (Chapman and Ward, 1996). This section develops a new approach that minimizes the total deviation from the ideal point to determine each criterion's weight. A numerical example is presented to show this approach's effectiveness in providing a complete ranking of all projects.

4.12 Model for the optimization in 3Ps

4.12.1 LP modelling

The division of equity financing between private partners and public agencies determines the sharing of project profit streams and affects 3Ps projects' successful delivery. An enhanced linear programming (ELP) model is developed to help public agencies accomplish their objectives while keeping those objectives attractive to private investments.

First, it is assumed that a 3Ps project spans T years. Funding is secured, and the project starts at time $t = 0$. The following model notations are used in the remainder of this part:

C = construction cost

D = debt

E_1 = private equity

E_2 = public funds

i_A = rate of return for public agency

i_B = rate of return for debt holders

i_P = rate of return for private partner

γ = public opportunity-loss coefficient

R_t = revenue at time (t)

DS_t = debt service at time (t)

OM_t = operation and maintenance costs at time (t)

DSR_t = debt service reserved payment at time (t)

$P_{1(t)}$ = profit sharing for private partner at time (t)

$P_{2(t)}$ = profit sharing for public agency at time (t); and

$DSCR$ = debt service coverage ratio.

For maximizing public interest, we have:

$$\max\left(D - \sum_{t=0}^{T} \frac{DS_t}{\left(1+i_A\right)^t} \right) + \left(E_t - \sum_{t=0}^{T} \frac{P_{1(t)}}{\left(1+i_A\right)^t} \right) - \gamma * E_2 \tag{4.18}$$

Subject to

$$D * DSCR - \sum_{t=0}^{T} \frac{DS_t}{\left(1+i_B\right)^t} \leq 0 \tag{4.19}$$

(Debt capacity constraint   debt holder interest)

$$DS_t * DSCR - (R_t + DSR_t - OM_t) \leq 0$$

(4.20)

(Debt service constraint debt holder interest)

$$C - (D + E_1 + E_2) \leq 0 \quad \text{(Minimal project funds constraint)}$$

(4.21)

$$E_1 - \sum_{t=0}^{T} \frac{P_{1(t)}}{\left(1 + i_{P(min)}\right)^t} \leq 0$$

(4.22)

(Project attractiveness constraint – private interests)

$$\sum_{t=0}^{T} \frac{P_{1(t)}}{\left(1 + i_{P(max)}\right)^t} - E_1 \leq 0$$

(4.23)

(Capping private equity return – public and private interests)

$$P_{1(t)} \leq R_t - OM_t - DS_t \text{ (Payment priority constraint)}$$

(4.24)

$$D, DS, E_1, E_2, P_1, P_2 \geq 0 \text{ (Nonnegative constraint)}$$

(4.25)

The objective of the optimization is to maximize the benefits of 3Ps financing for the public agency. Three components are included in the objective function: debt financing benefits (costs), private equity financing benefits (costs), and opportunity costs associated with public funds. The model must satisfy several constraints. First, the debt capacity constraint defines the maximal amount of debt that a 3Ps project can secure. Financial rating companies (e.g., Fitch and Standard and Poors) rate project debts following associated project risk and profitability. The bond rating for similar projects could be used to determine the debt service coverage ratio, which, along with the projected project revenue stream, determines the project's debt capacity. Second, the debt holders require that the debt service be secured with higher priority from net revenue. A reserve fund may also be created to pay debt service. The reserve fund would be from either (1) initial public or private investments or (2) operation profit reserves from earlier years. Third, 3Ps financing must be able to cover project costs. Fourth, private partners' return rates must be large enough to attract private investments yet small enough to protect public interests. $IP_{(min)}$ and $i_{P(max)}$ indicate the low and high boundaries of private equities' return rate. Furthermore, profits to private partners must be paid after debt services and reserve.

$$\min\left(DS - D\right) + \left(P_1 - E_1\right) + \gamma * E_2$$

Subject to

$$D * DSCR - R \leq 0 \tag{4.26}$$

$$DS = \alpha * D$$

$$D + E_1 + E_2 \leq 0C$$

$$P_1 \geq \beta_{min} * E_1 \qquad (SLP)$$

$$P_1 \leq \beta_{max} * E_1$$

$$P_1 \leq R - OM - DS \tag{4.27}$$

$$DS, D, P_1, E_1, E_2 \geq 0$$

Three types of financing mechanisms in 3Ps projects are debt, private equity, and public funds. The difference between debt service and debt represents public costs through debt financing. If the expected revenue is less, then the debt capacity available from banks will decrease. This decrease occurs because banks that accept debt consider debt service coverage ratio when calculating the amount of debt. The debt service coverage ratio is calculated as revenue available for debt service over debt service. In such cases, the finance gap is arranged through equity financing, costlier than the debts. In return for equity investments, private partners take a large share of project profits, translating into high return rates. Hence, public agencies need to fill the financial gap with private capital in the meantime to ensure that the return to private partners is not unexpectedly high. $(P_1 - E_1)$ represents the cost of private equity financing.

A reduction in upfront public investments may be beneficial to public agencies. These reduced upfront investments leave more money in hand to be used for other new or renovation jobs. Using public funds in a 3Ps project, a public agency essentially gives up the opportunity to build other infrastructure that could bring economic and social benefits to the public. In the enhanced linear programming and single-objective linear programming models, a public opportunity loss coefficient γ calculates the opportunity loss due to public funds in 3Ps projects. Profit sharing for the public agency should also be incorporated into the coefficient γ. When $\gamma = 1$, the amount of benefit from 3Ps project operations derived from funds invested in the public's 3Ps project will equal the cost of opportunity lost from alternative infrastructure development. The inequality $\gamma \prec 1$ indicates that opportunity cost is less than the benefits from the 3Ps projects. The opposite is true when $\gamma \succ 1$. The higher the γ is, the larger the opportunity loss is. In both models $\gamma * E_2$ represents the total opportunity cost of public funds in a 3Ps project.

4.12.2 Solving the model

When one considers that most 3Ps projects span a few decades, uncertainty exists and should be incorporated into the model. The randomness of project revenue is primarily investigated in this research. Three major techniques – robust optimization, stochastic programming, and probabilistic programming – are available for uncertainty modeling. Probabilistic constraints are the best method to model 3Ps capital structure's uncertainty issue for the following reasons. First, the robust optimization method requires that none of the constraints can be violated, and therefore that method is overly pessimistic because it chooses the worst-case scenario. Second, the stochastic programming approach assumes that the distributional information is known. However, this approach is overly optimistic for realistic cases. Third, the probabilistic constraints model controls the overall probability of constraints. More importantly, the probabilistic programming model measures the risk quantitatively.

There are three techniques to solve probabilistic programming: scenario optimization, average sample approximation, and polynomial approximation. The first two methods will provide a feasible solution rather than the optimal solution. The reason is obvious in that the general probabilistic constraints are usually nonconvex. A polynomial approximation approach was used to transform the original model into a convex optimization with polynomial constraints. The resulting model is solvable by most interior-point solvers. The polynomial approximation approach is fundamentally based on the properties of the Bernstein polynomial. A detailed discussion on the *Bernstein polynomial*[1] appears in Philips (2003). Then, the single-objective linear programming models can be modified to the simplified probability programming (SPP) model as follows:

$$\min (\alpha - 1) D + (P_1 - E_1) + \gamma * E_2 \quad (SPP)$$

Subject to

$$D + E_1 + E_2 = C$$

$$P_1 \geq \beta_{min} * E_1$$

$$P_1 \leq \beta_{max} * E_1 \tag{4.28}$$

$$\Pr (P_1 \leq R - OM - DS) \geq p$$

$$DS, D, P_1, E_1, E_2 \geq 0$$

$$D, E_1, E_2 \leq C$$

Where p is the probability, and the two constraints with random parameter R are modified as a probability function. Given a p-value of 95%, this probability constraint requires that the equation be held at a 95% confidence level.

4.12.3 United States case study

The Alabama Department of Transportation received an unsolicited proposal to build a 23-mi highway, named US-231–I-10 connector (see Figure 4.11), to run between the Alabama–Florida border and Dothan, Alabama. This highway was proposed to provide a safer and more efficient road network to relieve traffic congestion. Dothan, also known as the Hub of the Wiregrass, is located about 100 mi from Montgomery, Alabama, and about 200 mi from the Alabama cities of Birmingham and Mobile. The proposed highway will connect Dothan with these major population centers currently served by the Interstate system. The proposed connector starts at US-231 in Dale County, northwest of Dothan, and will follow a southerly direction, pass through Geneva County, and finally merge with US-231 near the Alabama–Florida border. This alignment will allow traffic to bypass Dothan and thus help to relieve current congestion in the city. If this connector is extended another 20 mi, it will connect to I-10 in Florida. The connector could be part of the corridor that private developers proposed to connect Montgomery, Alabama, and Panama City, Florida. Figure 4.7 shows the alignment proposed for the connector. The preliminary traffic and revenue study report estimated the cost of constructing the connector highway to be $100 million (the numbers have been adjusted within reasonable limits to maintain the confidentiality of actual numbers associated with the project). It also estimated the expected revenue streams from each traffic pattern: a base case and an external–external (EE) boosted-trip-table case.

Three scenarios were developed from the base case. The worst-case scenario assumed that the toll revenue growth (which incorporated traffic growth and toll growth with inflation) would be 4.6% for 30 years. Under the average scenario, the toll revenue growth rate would be 4.6% for the first ten years, 8% for the next ten years, and 4.6% for the last ten years. Under the best-case scenario, the toll revenue would grow at 4.6% for the first ten years and 8% for the next 20 years. Three more scenarios were developed from the external–external boosted revenue streams. However, under some of these scenarios, the toll revenue could not provide enough debt to cover all project costs. Therefore, equity financing would be required for this project. The reasonable distribution of private equity and public funds would remain the major concern for the state agency.

Both the single-objective linear programming models and simplified probability programming models were used to determine the optimal allocation of equity investment for the I-10 connector project. Beta distribution was used to calculate the present worth of expected revenue under each scenario. Furthermore, a sensitivity analysis was conducted to evaluate the impact of uncertainty in the toll revenue and the opportunity loss coefficient. Data sets used on the base run appear in Table 4.7. The optimal private equity investments under the base case and the external–external boosted case would be $9.55 million and $11.76 million, respectively. Under the simplified probability programming optimization model, given revenue R following a normal distribution, the

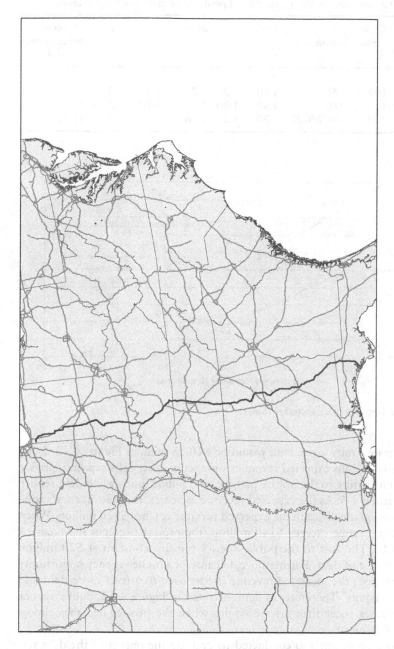

Figure 4.11 US-231–I-10 alignment (Source: ADOT).

Table 4.7 Data and results from a simplified probability programming analysis

Model	C ($ millions)	R ($ millions)	DSCR	α	β(min.)	β(max.)	γ		Optimal private equity ($ millions)
SLP									
Base case	100	65	1.50	1.20	1.36	2.10	2.00	9.55
EE boosted	100	80	1.50	1.20	1.36	2.10	2.00	11.76
SPP	100	N(80, 25)	1.50	1.20	1.36	2.10	2.00	10.25

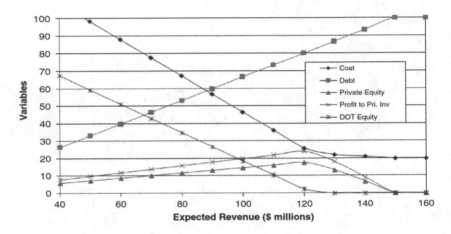

Figure 4.12 Impacts of expected revenue.

optimal private equity investment would be $10.25 million. Figure .4.12 shows the impact of various expected revenues on agency cost, debt capacity, private equity, profit sharing to the private partner, and public funds when debt service coverage ratio is 1.5. As the expected revenue increases, the state agency's total costs decrease gradually until the expected revenue reaches $120 million. When the expected revenue exceeds $140 million, the project becomes self-financed through debt. The cost to the public agency remains constant at $20 million to cover the cost of debt. Furthermore, because of a higher agency opportunity cost, in this case, the increased revenue is first used to attract more debt and then private equity. Therefore, as demonstrated in Figure 4.12, public agency equity decreases faster than private equity when the project becomes almost entirely self-financed.

Scenario analysis was also conducted to evaluate the impact of the debt service coverage ratio value on the equity structure. The debt service coverage ratio value was set from 1.35 to 1.75. The optimal equity structure was defined as the ratio of public equity (E_2) over private equity (E_1). When higher toll revenues are expected, the expectation of getting higher profits increases. The project then

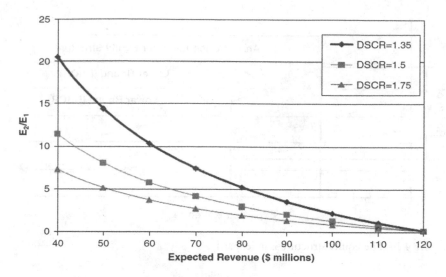

Figure 4.13 The equity structure changes.

becomes more interesting to private investors and the need for public funds is reduced. The equity structure at different debt service coverage ratio values is shown in Figure 4.13. Each curve on this graph is an optimal curve and can be used to obtain E_2 and E_1 by projecting the values of expected revenue from the x-axis to the optimal curves and then projecting them to the Y-axis. Given the debt service coverage ratio, the optimal equity structure, described as public fund over private equity (E_2/E_1), can be determined.

Considering the deterministic linear program model (SLP), the optimal equity structure for the I-10 connector project would be 3.0 under the base case and 5.8 under the external–external boosted case when the debt service coverage ratio is 1.5. However, if the debt service coverage ratio is varied over a range of 1.35 to 1.75 while the net revenues are assumed to be \$60 million, the equity structure ratio would range from 3.74 to 10.3. A value of zero for $E_2:E_1$ indicates that the optimal solution should have no public agency investment, but that does not mean that E_1 should also be zero. The amount of private equity investment can be determined only from Figure 4.5.

Figure 4.13 can be generalized for uncertain debt service coverage ratio values. When the debt service coverage ratio is uncertain over a range of low and high values, the optimal equity structure's upper and lower boundaries would be determined following the low and high debt service coverage ratio values. Because of the volatility of expected revenue flows, the optimal equity structure's left and right boundaries can also be determined. This determination will define an equity structure efficiency area, as shown in Figure 4.14. The area is dependent on debt service coverage ratio value and expected revenues. If the uncertainty in debt service coverage ratio is low, the optimal curves are much closer.

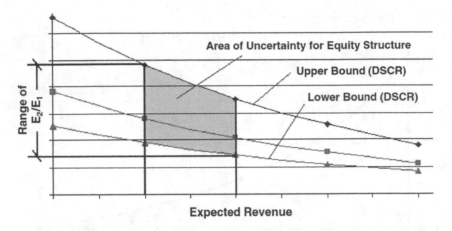

Figure 4.14 The equity structure is in jeopardy.

Similarly, if the expected revenue ranges within small intervals, then the efficiency space can be reduced. Under an extreme case where both debt service coverage ratio value and revenue flows are deterministic, the equity structure efficiency area will form a single point.

One of the model's major contributions is introducing the opportunity loss coefficient γ, which represents all opportunity costs associated with public investments (e.g., social benefits and transportation network improvement if the public investments are used in other projects). A sensitivity analysis indicates a threshold value γ that causes financing methods to switch from public funds to private equities. The threshold value indicates that alternative projects or other public works are highly beneficial for society or the transportation infrastructure network and should be given priority from the public agency's standpoint. The model presented here demonstrates a methodology that planners can adopt for balancing private and public interests in 3Ps contracts under uncertainty. The use of the public opportunity loss coefficient γ enables weighing public funds' social and external benefits, but the selection γ requires careful consideration.

Equity structure is critical to 3Ps project financing. In successfully delivering 3Ps projects, transportation agencies must carefully design the equity structure to attract private capital and simultaneously protect public interests. This chapter presents a model to help agencies maximize the benefits of 3Ps financing. The model includes the benefits and costs from debt and equity financing and allows users to incorporate opportunity loss into the evaluation. The case study discussed here shows that the optimal equity structure depends on three factors: expected toll revenue, debt service coverage ratio, and public opportunity loss coefficient. The research suggests that an optimal equity structure space could be defined under uncertainty. The model discussed here is especially useful for public agencies to (1) estimate the range of private equity investment, (2) identify the negotiation space for 3Ps contracts, and (3) determine the target equity structure in a 3Ps project.

The model does not include federal grants as a funding source. Suppose earmark funds or federal grants are available for 3Ps projects. In that case, users should consider the grants as a deductible item of total project costs financed through state agencies and private investors. Debt and equity financing could be from several sources and require different return rates, which would make the optimization problem much more complicated. Under this condition, users could either expand the model or consider the weighted average cost of capital. Furthermore, the analysis result is extremely sensitive to selecting parameters in the model; therefore, users should be cautious in selecting these parameters when designing the 3Ps financing plan.

4.13 Fighting corruption in public procurement

As public-sector operating budgets in transition and developing country economies continue to grow tighter, national and local government entities are increasingly looking to the private sector to provide goods and services to their constituents. Such goods and services are often acquired through a public procurement process whereby the government entity contracts with a private-sector enterprise to furnish a particular good or to provide a particular service for a fee subject to legal terms and conditions contained in a contract. The procurement process by which government entities award such contracts generally involves the following:

- The definition of the procurement requirement
- An estimated budget
- The solicitation of proposals
- The final award of a contract based on stated evaluation criteria and performance

In many cases, government entities seek private-sector providers to secure better-quality goods and services at a lower overall cost, that is, to obtain better value for money. Simultaneously, government entities are looking to streamline the procurement process to shorten delivery and performance times and reduce administrative costs. However, these objectives cannot generally be met unless contracts are awarded on a truly competitive basis under a system that has clear guidelines incorporating transparency, efficiency, economy, accountability, and fairness into the public procurement system as a whole. Because public procurement is one of the key areas where the public sector and the private sector interact financially, it is a prime candidate for corrupt activity, cronyism, favoritism, and outright bribery. Hence, public procurement has been targeted by various national, international, and multilateral anti-corruption initiatives as an area in need of reform.

In many countries, procurement reformers and anti-corruption advocates are rightfully taking a holistic view of approaching procurement reform and anti-corruption initiatives. Such holistic approaches include:

- Devising higher ethical standards for procurement officials
- Requiring asset disclosure for public officials of a certain rank or in a particular position
- Passing of freedom of information laws

Besides, audit and oversight institutions are being strengthened and given more authority – other initiatives train procurement personnel to achieve higher professional standards. Likewise, administrative and judicial dispute resolution institutions are being created to resolve problems arising from the solicitation process or contract performance and address corruption allegations. Finally, civil society organizations are strengthening their role as nongovernmental watchdogs. The private sector can help police the public procurement system by actively reporting procurement fraud, waste, and abuse and vigorously asserting contractual rights in appropriate fora.

While initiatives such as these are certainly positive developments, reformers need to consider the specific operational procurement process and the steps needed to develop a new procurement system. Although public procurement processes are fairly complex and can be implemented differently in various jurisdictions, the three main phases of the public procurement process are:

(I)   Procurement planning and budgeting
(II)   Procurement solicitation
(III) And contract award and performance

Corruption can arise in various forms in each of these separate phases of the procurement process.

In the procurement planning and budgeting phase, the government entity needs to determine what goods or services it wants to buy (the requirement) and how much it would like to spend (the budget). In both of these cases, there are opportunities for corruption. In determining the requirement, reports could be prepared that falsely justify current or future departmental needs, falsely inflate actual needs or falsely report damaged equipment to create an excess supply that could be used for corrupt purposes. The procurement requirements could also be written to favor or disfavor particular suppliers. Budgets could be set artificially high so that excess allocations can be stolen or diverted. Also, programmatic budgets could be devised so that there are overlapping budgetary allocations among separate organizations or departments that could likewise be applied in a corrupt manner (Liu et al., 2015; Liang et al., 2018; Zhang et al. 2019).

Procurement solicitation phase sets the main tasks of the request for proposals or tender documents and conduct the evaluation. The evaluation criteria in the request for proposals or tender documents could be drafted to favor a particular supplier or service provider or be drafted to emphasize a particular competitor's weaknesses. Later during the evaluation of the proposals or tenders, the evaluation criteria could be misapplied or otherwise further defined or amended after the proposal or tender receipt. During this phase, it is also possible that advance information could be provided to a particular favored supplier. Other techniques such as failing to solicit proposals or tenders from the competitors of a favored supplier, wrongfully restricting the tender pool, soliciting offerors known to be inferior to a favored supplier, simply misaddressing tender documents, accepting late proposals, or rejecting legitimate proposals are techniques that can be utilized to corrupt the procurement process.

Corruption opportunities also abound at the contract award and performance phase of the procurement process. For example, an offeror could propose an unrealistically low offer, hoping that procurement officials will allow amendments to increase costs after the contract is awarded. Likewise, a firm could offer exceptionally high-caliber products or less qualified personnel to meet a particular requirement and then, upon contract award, substitute inferior products or personnel. It is also possible to corruptly require sub-contractual relationships with favored suppliers. Furthermore, after the evaluation is complete, it is possible to award a contract that materially differs from the solicitation terms in terms of specifications, quantity, or delivery schedule. Oversight and reporting requirements may also be minimized, and in some cases, cost overruns can be corruptly explained away or falsely justified. Finally, supporting documentation could be intentionally lost or destroyed, making the detection and prosecution of corruption offenses difficult.

Whereas much holistic public procurement reform and anti-corruption activities are effective, considering the operational aspects of a procurement system can provide meaningful insights to public procurement reformers and anti-corruption advocates to achieve the goals mentioned above of promoting efficiency, economy, transparency, fairness, and accountability. For instance, *budgetary and financial controls should be operationally separate in the procurement planning phase*, and procurement requirements should be subject to internal and public scrutiny. Likewise, there should be clear guidelines in the proposal solicitation phase so that both procurement professionals and private-sector providers will understand their respective roles. *There should be an opportunity for the private sector to challenge the stated requirements and the solicitation evaluation result.* Finally, in the contract award and performance phase, strict financial controls and audit oversight are needed to protect the final phase of the procurement process's integrity.

These are but a few ideas as to how to address corruption in the context of procurement reform. Although no public procurement system will likely ever be fully free of all corruption, a system that promotes transparency, efficiency, economy, fairness, and accountability will be a system where corrupt activities will be more difficult to conceal and will be easier to punish administratively or criminally. More importantly, such a system will be more effective in providing a mechanism whereby the private sector can provide high-quality goods and services at a cost or price deemed fair and reasonable by the public sector. Understanding the procurement process's operational intricacies will help public procurement reformers and anti-corruption advocates contribute to this important endeavor.

4.14 3Ps Investment process

The 3Ps process is obvious as a branch of the broader public investment management process – that is, at some point, a project is chosen as a potential 3Ps and, after that, a 3Ps-specific process is followed. However, this branching can happen at different points in the public investment process. For instance, this could be:

After budgeting as a public investment project: as is the case in Australia and the Netherlands, procurement options (containing 3Ps) are evaluated only after a project has been confirmed and budgeted for as a public investment project. If the project is subsequently implemented as a 3Ps, then budget allocations are adjusted accordingly (Alkan et al., 2020).

After project appraisal and approval as a public investment: all public investment projects in Chile undergo a cost-benefit analysis via the National Planning Commission and meet a specified social return rate for public investments. 3Ps projects are also taken from this list.

After pre-feasibility or strategic options analysis: in the Republic of Korea, a potential 3Ps is identified after a pre-feasibility evaluation and a detailed project appraisal (such as technical feasibility studies or cost-benefit analysis). These are part of the 3Ps appraisal process. A similar strategy is followed in South Africa, where 3Ps implementation is part of an initial need's analysis and options assessment of a potential public investment project.

Well-defined 3Ps processes generally mirror public investment management processes – for example, requiring approvals by the same bodies, as described further in *Institutional Responsibilities: Review and Approval* (Irwin and Mokdad, 2010).

Many governments introduce criteria or checklists for 3Ps potential against which projects can be compared to support this screening process. 3Ps Potential Screening Factors in South Africa provide such a checklist from the South Africa 3Ps Manual (Baporikar, 2020). Similar criteria may also be used for more detailed appraisal, as described in *Assessing Value for Money of the PPP*. At the screening stage, the idea is to check if the criteria are likely to be met to proceed to the next development level.

4.14.1 The project selection problem

The project selection issue has received considerable attention in both decision analysis and supply chain management literature. It is becoming a fertile topic for study for operations research and management science disciplines. Ho et al. (2010) exhaustively review the individual and integrated decision-making strategies from 2000 to 2008 to aid the project selection problem. Chai et al. (2013) complementarily provide a systematic literature review of the decision-making theories assisting project selection from 2008 to 2012, which classifies the mentioned approaches into three categories: multiple-criteria decision-making

(MCDM) techniques, mathematical programming strategies, and artificial intelligence strategies.

Contemporary project management requires decision-makers to maintain strategic partnerships with few but reliable projects (Ho et al., 2010), which effectively reduce the project costs and improve the competitive benefits (Ghodsypour and O'Brien, 2001; Zhang et al., 2019). Therefore, besides the conventional price element, a promising project selection policy should also depend on a broad spectrum of qualitative and quantitative criteria such as quality, delivery, flexibility, and lead time, etc. (Chen et al., 2006). Dickson (1966) has identified 23 criteria to be considered during the project manager's process determining project selection.

The project selection issue examined in this chapter is described as follows. A set of candidate projects is evaluated in terms of criteria, with the involvement of a group of experts. Each expert has a specific preference for the ordering of criteria importance. In the presence of deterministic values for each project related to each criterion, each expert knows each project's evaluation results' lower and upper bounds. Therefore, an individual expert may produce interval evaluation values to measure each project's performance, such that an interval project selection matrix (ISSM) is formulated to support project evaluation and selection. Different experts may generate different intervals for specific projects. The interval formulation is motivated by the observation that in the domain of multiple criteria decision making, different weight elicitation methods may generate different weights even for the same issue, and it isn't easy to reach a consensus about exact weights (Lahdelma and Salminen, 2001). Evaluating a set of projects utilizing interval values is an important problem in decision analysis. This chapter aims to improve a sophisticated strategy for solving the interval project selection matrix above and provide a total ranking of the candidate projects. Although the large body of research on multicriteria project selection in the literature helps guide project managers to choose appropriate projects effectively, it is important to understand the impact of interval values on project evaluation and selection. To the best of our knowledge, the extant literature has left this interesting and important topic largely unexplored. The present study fills this gap by first formulating the interval project selection matrix (ISSM) and applying stochastic multicriteria acceptability analysis (SMAA-2) to provide a holistic candidate project rank. Such investigation sheds much-needed light on potential incentives and directions for academic, managerial, and policy-related implications.

Pioneered by Lahdelma et al. (1998), stochastic multicriteria acceptability analysis is a method intended to aid multiple criteria decision making with multiple experts in cases where little or no weight information is available. The criteria values are uncertain. It does not need the experts to describe their input data precisely or implicitly. It provides several meaningful and useful indices, containing an acceptability index for each alternative measuring the variety of input data that give each alternative the best-ranking position, central weight describing an expert's preferences supporting an alternative, and confidence factor representing the reliability of the analysis. Lahdelma and Salminen (2001) extend stochastic multicriteria acceptability analysis by considering all ranks and provide holistic SMAA-2 analysis to identify good compromise alternatives. For the issues with ordinal criteria information, Lahdelma et al. (2002) propose a new SMAA-O method. Durbach (2006) proposes an stochastic multicriteria acceptability analysis using achievement functions (SMAA-A) for a discrete-choice decision investigating what combinations of aspirations are necessary to make each alternative the preferred one. Lahdelma and Salminen (2006a) develop cross confidence factors based upon calculating confidence factors for alternatives using others' central weights. Lahdelma and Salminen (2006b) combine DEA and SMAA-2 to evaluate multicriteria alternatives. Lahdelma and Salminen (2009) develop the SMAA-P method that combines the piecewise linear difference functions of prospect theory with stochastic multicriteria acceptability analysis. Lahdelma et al. (2006, 2009) present and compare simulation and multivariate Gaussian distribution methods to treat the uncertainty and dependency information of the SMAA-2 multiple criteria decision making. Tervonen and Lahdelma (2007) present efficient methods for performing the computations through Monte Carlo simulation, evaluate the complexity, and evaluate the improved algorithms' accuracy. Corrente et al. (2014) integrate stochastic multicriteria acceptability analysis and Preference Ranking Organization METHod for Enrichment Evaluation (PROMETHEE) to explore the parameters compatible with the decision-makers' preferred information. Angilella et al. (2015) and Angilella et al. (2016) combine the Choquet integral preference model with stochastic multicriteria acceptability analysis to obtain strong recommendations and robust ordinal regression, respectively. Durbach and Calder (2016) investigate the context where decision-makers are unable or unwilling to evaluate trade-off information precisely in stochastic multicriteria acceptability analysis.

Besides the method development on stochastic multicriteria acceptability analysis, there exist important application papers in the literature: facility location (Lahdelma et al., 2002), forest planning (Kangas et al., 2006), elevator planning (Tervonen et al., 2008), descriptive multi-attribute choice model (Durbach, 2009a), estimation of a satisficing model of choice (Durbach, 2009b), Data envelopment analysis (DEA) cross efficiency aggregation (Yang et al., 2012), mutual funds' performance assessment (Babalos et al., 2012), and project portfolio optimization (Yang et al., 2015).

The main contribution of this chapter is summarized as follows. First, we formulate an interval project selection matrix to describe the project selection problem, in which each expert has specific but uncertain evaluation results on a set of candidate projects. Therefore, the project selection problem with interval values is deemed a stochastic optimization issue. Second, SMAA-2 is introduced via the concepts of rank acceptability index, central weight vector, and confidence element. Third, we apply SMAA-2 to the project selection problem with interval data and propose a holistic rank candidate project. Even though the classical project selection problem has been widely discussed in the context, such investigation in this study is completely new and of both academic and practical significance and value.

The remainder of this chapter is organized as follows. Section 4.14.2 presents the problem description. Section 4.15 introduces SMAA-2 and some related important indices. Section 4.16 applies SMAA-2 to solve the project selection with interval inputs. Section 4.17 concludes this study and proposes some future directions.

4.14.2 How to explain the Problem?

The project selection problem studied in this chapter is modeled as follows. A set of I candidate projects is evaluated in terms of J criteria, with a committee of K experts' involvement. All criteria are assumed to benefit. Concerning the cost-type criteria, we may take the transformation of negativity or reciprocal. Therefore, the basic framework of the multicriteria project selection problem is depicted by a decision matrix $G_{IJ} = \left[x_{ij} \right]_{IJ}$:

$$G_{IJ} = \begin{bmatrix} x_{11} & x_{12} & \cdots & x_{1J} \\ x_{21} & x_{22} & \cdots & x_{2J} \\ \vdots & \vdots & \vdots & \vdots \\ x_{I1} & x_{I2} & \cdots & x_{IJ} \end{bmatrix} \tag{4.29}$$

where $x_{ij}, x_{ij} \in [0,1], i = 1,2,...,I, j = 1,2,..., J$ are the exact values for all experts, and they have been normalized to delete the magnitude of data. The evaluation score of a project is calculated by the weighted sum of criteria measures concerning the mentioned project, that is,

$$S_i = \sum_{j=1}^{J} x_{ij} w_{ij}, i = 1, 2, ..., I \tag{4.30}$$

where w_{ij} are the weights of criterion j associated with the project i, and $\sum_{j=1}^{J} w_{ij} = 1, w_{ij} \geq 0$.

Each expert $k, k = 1, 2, ..., K$ is identified by a specific preference on the sequence of criteria. Without loss of generality, we assume that the criteria are arranged in descending order of importance for a typical expert $w_{i1}^k \geq w_{i2}^k \geq \cdots \geq w_{ij}^k$. This sequence changes across different experts. Therefore, a certain expert $k, k = 1, 2, ..., K$ may formulate the following mathematical model to aggregate the most favorable performance for each project i:

$$US_i^k = \max \sum_{j=1}^{J} x_{ij} w_{ij}^k$$

$$s.t. w_{i1}^k \geq w_{i2}^k \geq \cdots \geq w_{ij}^k, i = 1, 2, ..., I \tag{4.31}$$

$$\sum_{j=1}^{J} w_{ij}^k = 1, w_{ij}^k \geq 0, i = 1, 2, ..., I, k = 1, 2, ..., K.$$

Theorem 1 (Ng, 2008). The optimal score of projects i derived from Equation 4.31 is $\max\limits_{j=1,2,\ldots,J}\left\{\dfrac{1}{j}\sum\limits_{t=1}^{j}x_{it}\right\}$.

Proof. After denoting $v_{ij}^{k}=w_{ij}^{k}-w_{i(j+1)}^{k}\geq 0, j=1,2,\ldots,J-1, v_{iJ}^{k}=w_{iJ}^{k}\geq 0$, we obtain

$$\sum_{j=1}^{J}w_{ij}^{k}=\left(w_{i1}^{k}-w_{i2}^{k}\right)+2\left(w_{i2}^{k}-w_{i3}^{k}\right)+\cdots+(J-1)\left(w_{i(J-1)}^{k}-w_{iJ}^{k}\right)+J\left(w_{iJ}^{k}\right)$$

$$=v_{i1}^{k}+2v_{i2}^{k}+\cdots+Jv_{iJ}^{k} \tag{4.32}$$

$$=\sum_{j=1}^{J}jv_{ij}^{k}$$

$$=1.$$

We also incorporate $\varphi_{ij}^{k}=\sum\limits_{t=1}^{j}x_{it}$ and then have

$$\sum_{j=1}^{J}x_{ij}w_{ij}^{k}=x_{i1}w_{i1}^{k}+x_{i2}w_{i2}^{k}+\cdots+x_{iJ}w_{iJ}^{k}$$

$$=\left[\left(w_{i1}^{k}-w_{i2}^{k}\right)x_{i1}\right]+\left[\left(w_{i2}^{k}-w_{i3}^{k}\right)\left(x_{i1}+x_{i2}\right)\right]$$

$$+\cdots+\left[\left(w_{i(J-1)}^{k}-w_{iJ}^{k}\right)\left(x_{i1}+x_{i2}+\cdots+x_{i(J-1)}\right)\right] \tag{4.33}$$

$$+\left[w_{iJ}^{k}\left(x_{i1}+x_{i2}+\cdots+x_{iJ}\right)\right]$$

$$=v_{i1}^{k}\varphi_{i1}^{k}+v_{i2}^{k}\varphi_{i2}^{k}+\cdots+v_{iJ}^{k}\varphi_{iJ}^{k}$$

$$=\sum_{j=1}^{J}v_{ij}^{k}\varphi_{ij}^{k}.$$

Therefore, Equation 4.31 is equivalent to the following formulation:

$$US_{i}^{k}=\max\sum_{j=1}^{J}v_{ij}^{k}\varphi_{ij}^{k}$$

$$s.t.\sum_{j=1}^{J}jv_{ij}^{k}=1 \tag{4.34}$$

$$v_{ij}^{k}\geq 0, j=1,2,\ldots,J.$$

The second stage of Equation 4.34 is

$$\min z_i^k$$

$$s.t. \ z_i^k \geq \frac{1}{j}\varphi_{ij}^k \tag{4.35}$$

The optimal objective value of Equation 4.35 is obtained at the point that $z_i^k = \max\limits_{j=1,2,\ldots,J}\left\{\frac{1}{j}\varphi_{ij}^k\right\}$; the optimal objective value of Equation 4.31 is in terms

of $US_i^k = \max\limits_{j=1,2,\ldots,J}\left\{\frac{1}{j}\sum\limits_{t=1}^{j}x_{it}\right\}$. These are the most favorable evaluation values

determined by the expert k for project i, with the decision matrix's given input, Equation 4.29. Given the determined sequence of criteria provided by a typical expert, Equation 4.31 is easy to understand and simple to apply and can be effectively solved without the elicitation of the exact values of weights.

Similarly, it is also necessary to consider the least favorable evaluation scores by an expert k for the project i. Therefore, an analogous mathematical model is presented below:

$$LS_i^k = \min \sum_{j=1}^{J} x_{ij}w_{ij}^k$$

$$s.t. \ w_{i1}^k \geq w_{i2}^k \geq \cdots \geq w_{ij}^k, i = 1,2,\ldots,I \tag{4.36}$$

$$\sum_{j=1}^{J} w_{ij}^k = 1, w_{ij}^k \geq 0, i = 1,2,\ldots,I, k = 1,2,\ldots,K.$$

Theorem 2. The optimal score of projects i derived from Equation 4.36 is

$$\min\limits_{j=1,2,\ldots,J}\left\{\frac{1}{j}\sum\limits_{t=1}^{j}x_{it}\right\}.$$

On the strength of the obtained least and most favorable evaluation scores for the project i by an expert k, we formulate an interval project selection matrix $\Omega_{IK} = \left(\left[LS_i^k, US_i^k\right]\right)_{IK}$ that describes each expert's uncertain judgment on each project. Reasonable evaluation of the project i by an expert k should lie in $\left[LS_i^k, US_i^k\right]$:

$$
\Omega_{IK} = \begin{bmatrix}
\left[LS_1^1, US_1^1 \right] & \left[LS_1^2, US_1^2 \right] & \cdots & \left[LS_1^K, US_1^K \right] \\
\left[LS_2^1, US_2^1 \right] & \left[LS_2^2, US_2^2 \right] & \cdots & \left[LS_2^K, US_2^K \right] \\
\vdots & \vdots & \vdots & \vdots \\
\left[LS_I^1, US_I^1 \right] & \left[LS_I^2, US_I^2 \right] & \cdots & \left[LS_I^K, US_I^K \right]
\end{bmatrix}. \tag{4.37}
$$

Consistent with Yang et al. (2012), the derived interval project selection matrix can be viewed as a stochastic multiple criteria decision making problem. In the following section, we briefly introduce the SMAA-2 method proposed by Lahdelma and Salminen (2001), which effectively solves this stochastic multiple criteria decision making problem by providing a holistic rank of all alternatives.

4.15 Stochastic multicriteria acceptability analysis

Stochastic multicriteria acceptability analysis represents a family of methods for assisting multiple criteria decision making with uncertain, imprecise, or partially missing input data. The rationale behind stochastic multicriteria acceptability analysis is exploring the weight space to describe the preferences that make each alternative the most preferred one or grant a certain ranking position for a specific alternative. Lahdelma et al. (1998) initiate searching on this topic and propose rank acceptability index, central weight vector, and confidence factor for all alternatives. Lahdelma and Salminen (2001) extend the original stochastic multicriteria acceptability analysis method by considering all ranks in the analysis and provide a more holistic SMAA-2 analysis to identify good compromise alternatives graphically.

4.15.1 Preliminaries

In line with the interval project selection matrix introduced in Section 4.14.1, we consider that a committee of K experts has a set of I projects to be evaluated and selected. Neither expert-specific evaluation values nor weights are precisely known. We assume that a real-value utility function can represent the decision-maker's preferences across all experts' evaluations $g(i, w), i = \{1, 2, ..., I\}$. The weight vector w quantifies the decision-maker's subjective preferences across experts' judgments. Moreover, the uncertain evaluation values from experts on projects are represented by stochastic variables ξ_{ik} with assumed or estimated density function $f(\xi)$ in the space $X \subseteq \Re^{I \times K}$. Besides, the unknown weight vector is represented by a weight distribution with density function $f(w)$ in the set of feasible weights defined as

$$
W = \left\{ w \subseteq \Re^K : \sum_{k=1}^{K} w_k = 1, w_k \geq 0 \right\}. \tag{4.38}
$$

The total absence of weight vector information is represented in the '*Bayesian*' spirit by a uniform weight distribution in W, i.e., $f(w) = \dfrac{1}{Vol(\mathrm{W})} = \dfrac{(K-1)!}{\sqrt{K}}$.

The utility function is then used to map the stochastic experts' evaluation values and weight distributions into utility distributions $g(\xi_i, w)$ based upon the above descriptions.

We define a ranking function denoting the rank of each project as an integer from the best rank $(= 1)$ to the worst rank $(= I)$ as follows:

$$rank\,(\xi_i, w) = 1 + \sum_I \rho(g(\xi_l, w) > g(\xi_i, w)) \tag{4.39}$$

where $\rho(\text{true}) = 1$ and $\rho(\text{false}) = 0$.

The SMAA-2 method relies on analyzing the sets of favorable rank weights $W_i^r(\xi)$ defined as

$$W_i^r(\xi) = \{w \in W : rank(\xi_i, w) = r\} \tag{4.40}$$

in which a weight $w \in W_i^r(\xi)$ guarantees that alternative ξ_i obtains rank r.

4.15.2 Indexes

This subsection introduces several useful indexes proposed by the SMAA-2 method. The first one is the rank acceptability index, b_i^r, explained as the expected volume of the set of favorable rank weights. More specifically, b_i^r measures the variety of different valuations that grant alternative ξ_i rank r, which is computed by

$$b_i^r = \int_X f(\xi) \int_{W_i^r(\xi)} f(w)dwd\xi. \tag{4.41}$$

The rank acceptability index b_i^r belongs to the interval $[0,1]$. At the same time, $b_i^r = 0$ shows that the alternative ξ_i never reaches rank r and $b_i^r = 1$ represents that the alternative ξ_i always obtains rank r, neglecting the impact of the choice of weights. Furthermore, the rank acceptability index can be employed directly in the multicriteria evaluation of the alternatives. For large-scale problems, we develop an iterative process as below, in which the n best ranks (nbr) acceptabilities are analyzed at each interaction n:

$$a_i^n = \sum_{r=1}^{n} b_i^r. \tag{4.42}$$

The nbr-acceptability a_i^n is a measure of the different preferences that grant alternative ξ_i any of the n best ranks. This analysis proceeds until one or more alternatives obtain a sufficient majority of the weights.

The weight space concerning the n best rank associated with an alternative can be depicted by the concept of central *nbr* weight vector w_i^n as below:

$$w_i^n = \int\limits_{X} f(\xi) \sum_{r=1}^{n} \int\limits_{W_i^r(\xi)} f(w)\,w\,dw\,d\xi \,/\, a_i^n. \qquad (4.43)$$

Considering the given weight distribution, the central *nbr* weight vector is the best single vector representation for a decision-maker's preferences who assign an alternative any rank from 1 to n.

The third proposed index is the *nbr* confidence factor p_i^n, which is defined as the probability that the alternative reaches any rank from 1 to n if the central *nbr* weight vector is determined and computed by

$$p_i^n = \int\limits_{\xi:rank(\xi_i,w_i^n)} f(\xi)\,d\xi. \qquad (4.44)$$

More detailed knowledge about these indices can be found in Lahdelma and Salminen (2001). The manual on implementing stochastic multicriteria acceptability analysis in practice is provided by Tervonen and Lahdelma (2007).

4.15.3 Holistic analysis for the rank acceptabilities

On the strength of the acceptabilities for the above rank, following step is to develop a complementary approach that combines the rank acceptabilities into holistic acceptability indices associated with all alternatives as below:

$$a_i^h = \sum_{r=1}^{I} \alpha^r b_i^r \qquad (4.45)$$

where α^r are described as meta-weights for constructing holistic acceptability indices and satisfy $1 = \alpha^1 \geq \alpha^2 \geq \cdots \geq \alpha^I \geq 0$.

The elicitation of so-called meta-weights is essential in the weight determination process for a lexicographic decision problem, which reasonably assigns the largest value to α^1, and the least value to α^I. As for assigning weights to ranks, Barron and Barrett (1996) introduce three mechanisms: rank-sum approach, i.e., $\alpha^r(RS) = \dfrac{2(I+1-r)}{I(I+1)}, r=1,2,...,I$ reciprocal of the ranks approach,

i.e., $\alpha^r(RR) = \dfrac{1/r}{\sum\limits_{r=1}^{I} 1/r}, r=1,2,...,I$, and rank-order centroid approach, i.e.,

$\alpha^r(ROC) = \dfrac{1}{I}\sum\limits_{r=1}^{I}\dfrac{1}{r}, r=1,2,...,I$. We use ROC to determine $\alpha^r, r=1,2,...,I$

because they are more accurate, straightforward, and efficacious and provide an appropriate implementation tool (Barron and Barrett, 1996).

The holistic evaluation of rank acceptability indices generates an overall measure of the acceptability of all alternatives. This is helpful to rank and sort alternatives effectively.

4.15.4 *Numerical instance*

To apply SMAA-2 to solve the project selection problem, we draw the data from the multiple-criteria project selection problem studied by Xia and Wu (2007). Three criteria, namely, price, quality, and service, are rated using the three-point scale, i.e., 1, 2, and 3, which indicate 'low,' 'middle,' and 'high' for the price criterion, and 'good,' 'middle,' and 'poor' for quality and service criteria. The problem is to select 5 out of 14 candidate projects, with a committee of 6 experts. Each expert has a specific preference on the criteria importance, i.e., price-quality $\succ$ service, price $\succ$ service $\succ$ quality, quality $\succ$ price $\succ$ service, quality $\succ$ service $\succ$ price, service $\succ$ price $\succ$ quality, and service $\succ$ quality $\succ$ price, which are denoted by notations '1,' '2,' '3,' '4,' '5,' and '6,' respectively (see Table 4.8).

The interval project selection matrix $\Omega_{IK} = \left(\left[LS_i^k, US_i^k \right] \right)_{IK}$ is obtained by Equations 4.31 and 4.36, in which the interval evaluations on all projects by all experts are reported in Table 4.9.

Furthermore, the meta-weights to formulate the holistic acceptability indices are

$$\alpha^{12} = (1.00, 0.69, 0.54, 0.44, 0.36, 0.30, 0.25, 0.20, 0.16, 0.13, 0.10,$$
$$0.07, 0.05, 0.02) \tag{4.46}$$

Table 4.8 Data for project selection

Project	Price	Quality	Service	Price (norm)	Quality (norm)	Service (norm)
1	2	1	1	0.0600	0.0370	0.0400
2	3	1	1	0.0400	0.0370	0.0400
3	1	2	2	0.1200	0.0741	0.0800
4	2	2	2	0.0600	0.0741	0.0800
5	3	2	1	0.0400	0.0741	0.0400
6	1	2	3	0.1200	0.0741	0.1200
7	1	3	1	0.1200	0.1111	0.0400
8	1	1	3	0.1200	0.0370	0.1200
9	2	2	1	0.0600	0.0741	0.0400
10	2	2	3	0.0600	0.0741	0.1200
11	3	3	1	0.0400	0.1111	0.0400
12	3	2	2	0.0400	0.0741	0.0800
13	2	3	1	0.0600	0.1111	0.0400
14	2	1	3	0.0600	0.0370	0.1200

Table 4.9 Interval project selection matrix

Project	Expert					
	1	2	3	4	5	6
1	[0.0457, 0.0600]	[0.0457, 0.0600]	[0.0370, 0.0485]	[0.0370, 0.0457]	[0.0400, 0.0500]	[0.0385, 0.0457]
2	[0.0385, 0.0400]	[0.0390, 0.0400]	[0.0370, 0.0390]	[0.0370, 0.0390]	[0.0390, 0.0400]	[0.0385, 0.0400]
3	[0.0914, 0.1200]	[0.0914, 0.1200]	[0.0741, 0.0970]	[0.0741, 0.0914]	[0.0800, 0.1000]	[0.0770, 0.0914]
4	[0.0600, 0.0714]	[0.0600, 0.0714]	[0.0670, 0.0741]	[0.0714, 0.0770]	[0.0700, 0.0800]	[0.0714, 0.0800]
5	[0.0400, 0.0570]	[0.0400, 0.0514]	[0.0514, 0.0741]	[0.0514, 0.0741]	[0.0400, 0.0514]	[0.0400, 0.0570]
6	[0.0970, 0.1200]	[0.1047, 0.1200]	[0.0741, 0.1047]	[0.0741, 0.1047]	[0.1047, 0.1200]	[0.0970, 0.1200]
7	[0.0904, 0.1200]	[0.0800, 0.1200]	[0.0904, 0.1156]	[0.0756, 0.1111]	[0.0400, 0.0904]	[0.0400, 0.0904]
8	[0.0785, 0.1200]	[0.0923, 0.1200]	[0.0370, 0.0923]	[0.0370, 0.0923]	[0.0923, 0.1200]	[0.0785, 0.1200]
9	[0.0580, 0.0670]	[0.0500, 0.0600]	[0.0580, 0.0741]	[0.0570, 0.0741]	[0.0400, 0.0580]	[0.0400, 0.0580]
10	[0.0600, 0.0847]	[0.0600, 0.0900]	[0.0637, 0.0847]	[0.0741, 0.0970]	[0.0847, 0.1200]	[0.0847, 0.1200]
11	[0.0400, 0.0756]	[0.0400, 0.0637]	[0.0637, 0.1111]	[0.0637, 0.1111]	[0.0400, 0.0637]	[0.0400, 0.0756]
12	[0.0400, 0.0647]	[0.0400, 0.0647]	[0.0570, 0.0741]	[0.0647, 0.0770]	[0.0600, 0.0800]	[0.0647, 0.0800]
13	[0.0600, 0.0856]	[0.0500, 0.0704]	[0.0704, 0.1111]	[0.0704, 0.1111]	[0.0400, 0.0704]	[0.0400, 0.0756]
14	[0.0485, 0.0723]	[0.0600, 0.0900]	[0.0370, 0.0723]	[0.0370, 0.0785]	[0.0723, 0.1200]	[0.0723, 0.1200]

Table 4.10 Normal distribution of holistic acceptability

Project	b^1	b^2	b^3	b^4	b^5	b^6	b^7	b^8	b^9	b^{10}	b^{11}	b^{12}	b^{13}	b^{14}	a^b
1	0.00	0.00	0.00	0.00	0.00	0.00	0.00	0.00	0.00	0.00	0.00	0.01	0.35	0.64	0.0307
2	0.01	0.01	0.00	0.01	0.02	0.03	0.03	0.02	0.04	0.04	0.08	0.11	0.27	0.33	0.0960
3	0.00	0.25	0.46	0.23	0.05	0.00	0.00	0.00	0.00	0.00	0.00	0.00	0.00	0.00	0.5392
4	0.00	0.00	0.00	0.00	0.00	0.09	0.47	0.38	0.07	0.00	0.00	0.00	0.00	0.00	0.2311
5	0.00	0.00	0.00	0.00	0.00	0.00	0.00	0.00	0.00	0.00	0.04	0.55	0.37	0.03	0.0607
6	0.91	0.08	0.01	0.00	0.00	0.00	0.00	0.00	0.00	0.00	0.00	0.00	0.00	0.00	0.9708
7	0.07	0.24	0.16	0.16	0.21	0.12	0.03	0.01	0.00	0.00	0.00	0.00	0.00	0.00	0.5128
8	0.03	0.24	0.18	0.26	0.17	0.06	0.05	0.02	0.01	0.00	0.00	0.00	0.00	0.00	0.5035
9	0.00	0.00	0.00	0.00	0.00	0.00	0.00	0.00	0.03	0.10	0.59	0.27	0.01	0.00	0.0962
10	0.02	0.17	0.17	0.26	0.31	0.09	0.01	0.00	0.00	0.00	0.10	0.00	0.00	0.00	0.4833
11	0.00	0.00	0.00	0.01	0.03	0.08	0.13	0.12	0.22	0.29	0.10	0.01	0.00	0.00	0.1798
12	0.00	0.00	0.00	0.00	0.00	0.00	0.01	0.13	0.31	0.39	0.14	0.03	0.00	0.00	0.1464
13	0.01	0.00	0.01	0.05	0.09	0.28	0.14	0.20	0.16	0.06	0.01	0.00	0.00	0.00	0.2630
14	0.00	0.01	0.01	0.02	0.12	0.25	0.13	0.12	0.16	0.11	0.05	0.03	0.00	0.00	0.2426

The SMAA-2 model can be effectively solved by the open-source software developed by Tervonen (2014, 2008).

4.15.5 Normal distribution

We consider that the interval data $\left[LS_i^k, US_i^k \right]$ are normally distributed, the mean and variance represented by $\mu_i^k = \dfrac{LS_i^k + US_i^k}{2}$ and $\left(\sigma^2 \right)_i^k = \dfrac{US_i^k - LS_i^k}{6}$, respectively. The results of the rank acceptability indices and the holistic acceptability indices derived according to SMAA-2 are shown in Table 4.10 and graphically reported in Figure 4.15.

Based upon the holistic acceptability indices in Table 2.10, we obtain a full and comprehensive rank of all projects: $6 \succ 3 \succ 7 \succ 8 \succ 10 \succ 13 \succ 14 \succ 4 \succ 11 \succ 12 \succ 9 \succ 2 \succ 5 \succ 1$ – the selected projects are projects 6, 3, 7, 8, and 10. The most profitable project is project 6, whose holistic rank index is 97.08% and first rank support is 91% possibility. In contrast, the least profitable project is project 1, the holistic rank index, and the last rank support of 3.07% and 64% of the possibility.

4.15.6 Uniform distribution

We alternatively assume that the interval data are uniformly distributed. With such assumptions, we report the holistic acceptability indices and the rank acceptability indices in Table 4.11 and Figure 4.16.

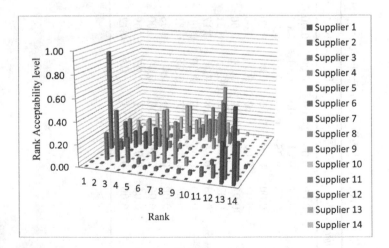

Figure 4.15 Rank acceptability indices (normal distribution).

Table 4.11 Uniform distribution of holistic acceptability

Project	b¹	b²	b³	b⁴	b⁵	b⁶	b⁷	b⁸	b⁹	b¹⁰	b¹¹	b¹²	b¹³	b¹⁴	aᵇ
1	0.00	0.00	0.00	0.00	0.00	0.00	0.00	0.00	0.00	0.00	0.00	0.03	0.55	0.41	0.0362
2	0.01	0.00	0.00	0.00	0.01	0.01	0.01	0.01	0.02	0.02	0.04	0.07	0.23	0.56	0.0587
3	0.02	0.23	0.36	0.26	0.11	0.02	0.00	0.00	0.00	0.00	0.00	0.00	0.00	0.00	0.5321
4	0.00	0.00	0.00	0.00	0.01	0.16	0.43	0.32	0.08	0.01	0.00	0.00	0.20	0.00	0.2365
5	0.00	0.00	0.00	0.00	0.00	0.00	0.00	0.00	0.00	0.03	0.14	0.61	0.00	0.02	0.0708
6	0.82	0.13	0.04	0.01	0.00	0.00	0.00	0.00	0.00	0.00	0.00	0.00	0.00	0.00	0.9359
7	0.10	0.19	0.17	0.16	0.17	0.12	0.04	0.03	0.02	0.01	0.00	0.00	0.00	0.00	0.5103
8	0.03	0.23	0.19	0.20	0.15	0.08	0.05	0.03	0.02	0.01	0.01	0.01	0.00	0.00	0.4813
9	0.00	0.00	0.00	0.00	0.00	0.00	0.00	0.01	0.06	0.17	0.55	0.19	0.01	0.00	0.1025
10	0.02	0.17	0.18	0.24	0.28	0.09	0.02	0.00	0.00	0.00	0.00	0.00	0.00	0.00	0.4717
11	0.00	0.01	0.01	0.02	0.05	0.10	0.12	0.14	0.19	0.24	0.09	0.02	0.00	0.00	0.1995
12	0.00	0.00	0.00	0.00	0.00	0.01	0.05	0.18	0.34	0.31	0.10	0.02	0.00	0.00	0.1592
13	0.01	0.01	0.03	0.05	0.11	0.22	0.16	0.16	0.14	0.10	0.01	0.00	0.00	0.00	0.2687
14	0.00	0.01	0.03	0.05	0.12	0.20	0.13	0.13	0.13	0.10	0.06	0.04	0.01	0.00	0.2495

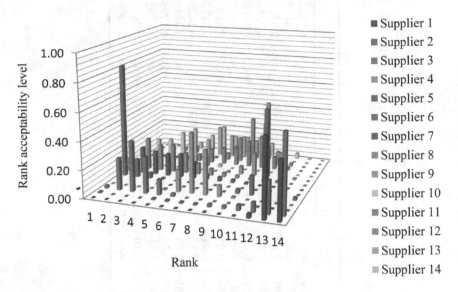

Figure 4.16 Rank acceptability indices (uniform distribution).

It is observed that the sequence of candidate projects using SMAA-2 under uniform distribution is: $6 \succ 3 \succ 7 \succ 8 \succ 10 \succ 13 \succ 14 \succ 4 \succ 11 \succ 12 \succ 9 \succ 5 \succ 2 \succ 1$, and the selected projects are projects 6, 3, 7, 8, and 10 as well. This sequence is mildly different from that derived from the norm distribution case. The only difference lies in the rank positions of projects 2 and 5. In detail, the most profitable project 6's holistic rank index is 93.59%, and first rank support is 82% of the possibility, both of which are lower than that of a normal distribution case. Meanwhile, the holistic rank index and last rank support the possibility of the least favorable supplier 1 are 3.62% and 41%, respectively.

In summary, SMAA-2 under both the normal distribution and uniform distribution assumptions may produce complete ranks with sufficient discrimination power among all alternatives; in that case, each expert has uncertain evaluations across all projects.

The multicriteria project selection problem with a group of experts has been widely explored in decision science and supply chain management literature. Given the exact input data, different experts may generate uncertain evaluation results for all projects. However, the extant literature has left this topic largely unexplored. This chapter is initially engaged in this tremendous wave by first formulating the interval values to optimize and then innovatively applying the SMAA-2 method to obtain an overall ranking of candidate projects. The interval data are assumed to be either normally or uniformly distributed in this study. A meta-weight scheme to derive holistic rank indices is elicited from the previous literature. A numerical example from the current work is reexamined to show the effectiveness of our approach.

This part of the research provides the decision-maker with more methodological options and enriches the project selection problem's theory and method. Future research should consider the determination of the uncertain sets for decision-making and investigate more practical distributions over the uncertainties.

4.16 Consequences

This section discusses theoretical and practical implications from the analysis of the developed model with the case study. The first theoretical implication is the values of private evaluative criteria and public evaluative criteria, whose values have been hardly accurate in previous studies. To evaluate these two kinds of evaluative criteria, this chapter introduces hesitant fuzzy set to describe the evaluative criteria values, and fully considers multi-criterion information given by the group. The method reveals new insight into the evaluative approaches on both private and public evaluative criteria of 3Ps infrastructure projects. Some characteristics of 3Ps infrastructure projects still need to be considered further with the evaluative values, especially for the matching decision cases, to satisfy the match-degree of participants, decrease matching risks, and produce an optimal matching scheme.

The second insight is selecting an optimal matching method by balancing private evaluative criteria and public evaluative criteria. Based on the values of private evaluative criteria and public evaluative criteria, the research uncovered the government's match-degree and the enterprise considering different evaluative criterion weights for 3Ps infrastructure projects. It reduces the best matching method by applying the min-max approach. The developed model provides a new perspective for selecting the best matching method for 3Ps infrastructure projects.

This chapter's practical implication is to assist DMs in selecting the most suitable matching method dynamically and objectively. The government's subjects can improve their sincerity level and enthusiasm for taking large risks, react to the dynamics of prices to dispel the misgivings of the subjects in the enterprise, and ensure the success of matching. Moreover, applying the proposed model to the example demonstrated that the best matching scheme of 3Ps infrastructure projects is to optimize matching partly from the perspective of subjects in the government and consider the matching satisfaction of the enterprise's subjects. It helps the DMs to make the best decision.

4.17 Conclusions

In the first part of this chapter, we developed a quantitative matching decision model of Multi-criteria Decision Making in 3Ps infrastructure projects based on hesitant fuzzy set by balancing the match-degree of the public and private sectors. The major contributions of the chapter are summarized as follows. First, to measure the match-degree of the two sides, this model, in which information on evaluative criterion weights is completely unknown, utilized hesitant fuzzy set to describe multi-criterion information given by a group. Second, the best decision method was established by maximizing each criterion's absolute deviation to determine the evaluative criterion weights objectively. The proposed matching decision model and its application are given by balancing the two parts' evaluative criteria. Third, discussion and implication reveal that the risks of land policy changes, legal policy changes, and exchange rate changes of the government's subjects can also affect the selection of the best matching decision method. As a result, the proposed matching decision approach expands a single value in existing Multi-criteria Decision Making evaluative values, which is closer to hesitant fuzzy and uncertain information in real matching decisions. Certain limitations need to be further studied in the future. The research still needs to be investigated and explored more extensively for the best matching plan, especially from the associated relationship between the evaluative criteria.

The evaluation for the critical issues involved in the early stages of 3Ps (i.e., *ex-ante* evaluation) has received limited attention. With this significant knowledge gap, this chapter reviews the normative literature on 3Ps. It conducts a comparative study between the approaches that are widely used in construction project evaluations (e.g., key performance indicators framework, quality-based excellence models, Balanced Scorecard, and Performance Prism). The critical comparison suggests that the Performance Prism is the most suitable performance measurement framework for *ex-ante* evaluation of infrastructure 3Ps, owing to its strong capabilities of fully capturing 3Ps characteristics (e.g., value for money, integration of multiple stakeholders, and dynamic and long-term development process). Based on this finding, a new performance measurement systems has been conceptualized for 3Ps.

This conceptually proposed performance measurement systems integrates a set of core indicators derived from the review of the normative literature. It can synergize with 3Ps *ex-ante* evaluation by providing public authorities and supporting agencies with critical insight into the balanced examination for the essential issues identified in the early stages of the projects. As this is the only conceptual chapter, the future study will focus on the validations of the developed CIs and the proposed performance measurement systems by applying a questionnaire survey and case study, respectively.

The multicriteria project selection problem with a group of experts has been widely explored in decision science and supply chain management literature. Given the exact input data, different experts may generate uncertain evaluation results for all projects. However, the extant literature has left this topic largely undiscovered. This chapter is initially engaged in this tremendous wave by first formulating the interval values to optimize and then innovatively apply the SMAA-2 method to overall rank overall candidate projects. The interval data are assumed to be either normally or uniformly distributed in this study. A meta-weight scheme to derive holistic rank indices is elicited from the previous literature. A numerical example from the current work is reexamined to show the effectiveness of our approach.

This chapter provides the decision-maker with more methodological options and enriches the project selection problem's theory and method. Future research should consider the determination of the uncertain sets for decision-making and investigate more practical distributions over the uncertainties.

Note

1 In the mathematical field of numerical analysis, a Bernstein polynomial, named after Sergei Natanovich Bernstein, is a polynomial in the Bernstein form, that is a linear combination of Bernstein basis polynomials.

  A numerically stable way to evaluate polynomials in Bernstein form is de Casteljau's algorithm.

  Polynomials in Bernstein form were first used by Bernstein in a constructive proof for the Stone–Weierstrass approximation theorem. With the advent of computer graphics, Bernstein polynomials, restricted to the interval [0, 1], became important in the form of Bézier curves.

5 Investment decision and corruption

5.1 Investment process in 3Ps

A 3Ps is a relationship between the public party and the private party established in a certain way so as to complete the investment and construction of public infrastructure together. The public party (usually the government) would guarantee the private party franchise right in exchange for the acceleration of the construction process and efficient management afterward (Zheng et al., 2018). Since the 3Ps model can combine both the public and private parties' advantage and improve the service quality on the premise of reducing construction investment, it has been widely used in foreign countries in the construction area since its first application in England in the 1990s.

3Ps projects are usually public infrastructure, which involves huge investment, exceptionally long construction periods, irreversibility, uncertainty, and competition. Thus, they are especially suitable for investigation with the real options method.

5.2 The properties of 3Ps

3Ps projects are separated into two parts: the establishment and the construction part, and the operation and the management part. The first part includes feasibility research, land acquisition, business invitation, plan design, construction, check, and acceptance at last. Based on completing the first section of the project, the operation and management section conducts the project and gains profit (Min et al., 2009).

During the construction phase, the private enterprise has a negative cash flow from the 3Ps project, so the project's NPV must be negative, corresponding to the option fee from the view of real options. After the option fee payment, project investors would gain franchise rights for land development, operation, and selling in a certain period. Those consequent decisions have considerable flexibility and uncertainty from the administration, economy, city planning, and government policy, so it isn't easy to evaluate these projects.

Because of all the uncertainties, private investors tend to develop the project by stages rather than put all money in at the very beginning, significantly reducing the risks since the investor can decide whether and when to take the further

DOI: 10.4324/9781003177258-5

step according to the market circumstances after finishing the first stage. Besides, the time for developing, transfer, and operating is fixed, so these projects are recognized as a European call option.

Under the option's view, 3Ps projects' total value contained two parts: net value and option value (Trigeorgis L, 1996). It is defined as Expanded net present value: Expanded net present value = traditional net present value + project' option value If Expanded net present value is positive, the project is accepted and otherwise not.

5.3 The Value of 3Ps

Based on Black-Scholes (B-S) option pricing algorithm (Black and Scholes, 1973), five major factors affect the option value (S_t, X, r, T, σ). To use the B–S model to analyze 3Ps projects, we need to distinguish the option factors contained in projects corresponding to financial option factors and explain as follows: V_t refers to the project's value at a particular time t; $M(V_t, t)$ refers to the real option value at time t; σ; refers to the volatility of the project's return rate; r refers to the risk-free interest rate; T is the projected completion time.

Based on the Black-Scholes algorithm, we can get an option pricing model for 3Ps projects as follows:

$$M(V_t, t) = V_t \cdot N(\widehat{d_1}) - Ie^{-r(T-t)} \cdot N(\widehat{d_2}) \tag{5.1}$$

In which:

$$\widehat{d_1} = \left[\ln\frac{V_t}{I} + \left(r + \frac{1}{2}\sigma^2 \right)(T-t) \right] \Big/ \sigma\sqrt{T-t}$$

$$\widehat{d_2} = \left[\ln\frac{V_t}{I} + \left(r + \frac{1}{2}\sigma^2 \right)(T-t) \right] \Big/ \sigma\sqrt{T-t}$$

$I = I_0 + A(1+i)^n \cdot I_0$ refers to capitalization; A refers to the loan; I refer to the interest rate; n refers to the loan term.

Since the Black-Scholes algorithm is more familiar with actual use, according to six basic hypotheses of the Black-Scholes algorithm (Black and Scholes, 1973; Mao and He, 2009), the 3Ps real option pricing model's major parameters are listed as follows:

- 3Ps project's value $-V_t$: V_t corresponds to the stock price S_t. We can use the continuous method to analyze 3Ps projects' price evolvement process, as the whole project's value obeys Brownian motion and will not suddenly go up and down. The project's price obeys normal distribution.
- The present value of the 3Ps project's investment $-I$: I corresponds to the option strike price X. In 3Ps, project investment usually contains

capitalization I_0 and financing money; financing money is mainly a construction loan and its interest during the construction period. We can use the present year's lending rate when calculating this parameter, and obviously, loan term n should coincide with the construction periods.

- The volatility of the 3Ps project's return rate $-\sigma$: σ corresponds to the volatility of the stock's return rate. It is the project's return rate's immediate standard error. We use historical data of similar projects to approximate the volatility for the actual project.

- The 3Ps project's franchise right term $-T$: T corresponds to the maturity date of the option contract. 3Ps projects' franchise right term is usually fixed in the contract between the government and the private enterprise to be viewed as constant. However, it might change according to the project circumstance.

- The 3Ps project's risk-free interest rate, r: under risk-neutral condition, all stocks' return rate is risk-free interest rate r, so we conclude that the primary option value is equal to maturity value discounted by risk-free rate. The risk-free rate is usually decided according to the *AAA* bond yield[1] rate curve, or we can use the last year's Treasury bill rate as the risk-free rate.

5.4 A Case from East Asia

In a south Asian countries Metro project for a line was first 3Ps in this country.. It has a total investment of 15.3 billion, of which the municipal government contributes 70% while the rest is left to the private enterprise which bid for the franchise right. The municipal government takes charge of the railway construction. The private enterprise invests in the vehicles and other facilities and runs Line 4 after its completion until it is transferred to the government for free when the 30 years' franchise term is over.

5.4.1 *Old Net Present Value*

In a private enterprise perception, the project's parameters are defined as follows: total investment 15.3b × 30% = 4.59b; self-capital 4.59b × 1/3 = 1.53 0.1 billion; loan: 4.59b × 2/3 = 3.06b; construction period: four years (2005–2008), and the money was put in averagely by year; franchise term: 30 years (2009–2038); lending rate: 6.12% in 2005; the expected after-tax revenue was 0.3b for the first year, and the growth rate was 5%; the expected operation cost was 0.1b, and the growth rate is also 5%; the loan was supposed to be repaid averagely during the first ten years after the metro was put into operation.

The project's discounted net cash flow value was listed in Table 5.1. We can get the discounted accumulative net cash flow (NPV):

$$NPV = \sum_{i=2005}^{2038} Net\ cash\ flow_i = -13440.2$$

Table 5.1 Changing the range of expanded NPV

The changing range of every factor	The corresponding changing rate of expanded NPV		
	Project value V	Volatility σ	Risk-free rate r
0.50	−1.14539	−0.02936	−0.13037
1.50	1.15632	0.02650	0.07174

Based on the calculation, the project's NPV was negative, so private enterprises should not invest in it.

5.4.2 Upgraded Net Present Value

According to this project, we get the parameters for the option pricing algorithm of 3Ps projects as follows: T was 34 years; total investment I was 42 billion; the volatility was estimated 9%, $\sigma^2 = 0.09$; the risk-free interest rate r was 5.76% (deposit rate for 5 years above); using the option pricing algorithm for 3Ps projects, Equation 5.1:

When $t = 0$:

$$\widehat{d_1} = \left[\ln\frac{V_t}{I} + \left(r + \frac{1}{2}\sigma^2\right)(T-t)\right] \Big/ \sigma\sqrt{T-t} = 1.9756$$

$$\widehat{d_2} = \left[\ln\frac{V_t}{I} + \left(r + \frac{1}{2}\sigma^2\right)(T-t)\right] \Big/ \sigma\sqrt{T-t} = 0.2263$$

Using the NORMSDIST () function,[2] we get;

$$N(\widehat{d_1}) = N(1.9756) = 0.9759$$

$$N(\widehat{d_2}) = N(0.2263) = 0.5895$$

And thus:

$$M(V_t, t) = V_t \cdot N(\widehat{d_1}) - Ie^{-r(T-t)} \cdot N(\widehat{d_2}) = 362543.7 \text{ million}$$

The upgraded Net Present Value = old Net Present Value + option value of the project, so we can calculate the expanded NPV=−13440.2+362543.7 = 349103.5 m>0. So, private enterprises should accept this project.

5.4.3 Sensitivity analysis[3]

We considered five parameters affecting the option value, project value V, the present value of investment I, volatility σ, time T, and risk-free interest rate r.

T and I are already certain, so we need to perform single-factor sensitivity analysis, respectively, on V, σ, and r.

Upgraded Net Present Value. The intervals of project value V, volatility σ, and risk-free rate r are respectively $[0.5V, 1.5V]$, $[0.5\sigma, 1.5\sigma]$, and $[0.5r, 1.5r]$. Using Excel for sensitivity analysis, we draw Figure 5.1.

The extreme point data in Figure 5.1 are listed in Table 5.1.

Based on Figure 5.1, we conclude that project value V is the most sensitive factor among the three factors, and the volatility σ is the least sensitive factor, with risk-free rate r in between. So, the calculation of the project value V is essential since it has a crucial influence on the final decision.

Comparing results by the old net percent value strategy and the new evaluation model, we see that the new algorithm's result fits reality. Thus, we conclude that the new algorithm, which considers the 3Ps project's option value, is more exact than the net percent value method.

5.5 The analysis of net present based on the number of actual positive internal rates of return

Considering that we are dealing with an investment project, the net percent value function must have at least one change of sign in the cash flow stream: the first cash flows will be negative (investment time), followed by cash flows that may change signs several times.

$$NPV(r) = C_0 + \frac{C_1}{1+r} + \frac{C_2}{(1+r)^2} + \dots + \frac{C_m}{(1+r)^m} + \frac{C_{m+1}}{(1+r)^{m+1}} + \dots + \frac{C_n}{(1+r)^{n}} \quad (5.2)$$

where C_i is negative if $I \le m$, and C_i might be positive or negative if $I \succ m$.

Several important points can be recognized when this function is drawn on a diagram:

Theorem 1. As the value of r reaches $-\infty$:

$$\lim_{r \to -\infty} NPV(r) = C_0 + \frac{C_1}{-\infty} + \frac{C_2}{(-\infty)^2} + \dots + \frac{C_n}{(-\infty)^n} = C_0. \quad (5.3)$$

But since this is an investment project, $C_0 \prec 0$.

The function has a vertical asymptote in $r = -1$. It is available, also, to recognize what occurs immediately to the right and the left of this asymptote:

Theorem 2. Left of the asymptote:

$$\lim_{r \to -1-\varepsilon} NPV(r) = C_0 + \frac{C_1}{-\varepsilon} + \frac{C_2}{(-\varepsilon)^2} + \dots + \frac{C_n}{(-\varepsilon)^n} \quad (5.4)$$

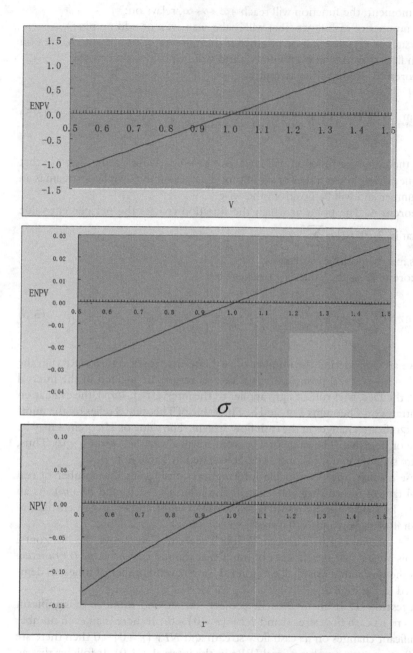

Figure 5.1 Expanded net present value's changing pattern from *V*, σ, *r*.

At this moment, the function will reach $+\infty \leftrightarrow -\infty$, relays on:

The number of cash flows (that will make n an even or odd number).

The sign of C_n (that will be negative if the number of significant changes on the cash flow stream is an even number and will be positive otherwise).

Theorem 3. Right of the asymptote:

$$\lim_{r \to -1+\varepsilon} NPV(r) = C_0 + \frac{C_1}{+\varepsilon} + \frac{C_2}{(+\varepsilon)^2} + \dots + \frac{C_n}{(+\varepsilon)^n} \tag{5.5}$$

At this theorem, the function will reach $+\infty \leftrightarrow -\infty$, relying on the sign C_n; that will be negative if the number of significant changes on the cash flow stream is an even number and will be positive otherwise.

Theorem 4. The point at which the function crosses the net percent value axis, that is, $NPV(0) = \sum_{i=0}^{n} C_i$.

This may be positive or negative.

Theorem 5. As the value of r reaches $+\infty$:

$$\lim_{r \to +\infty} NPV(r) = C_0 + \frac{C_1}{+\infty} + \frac{C_2}{(+\infty)^2} + \dots + \frac{C_n}{(+\infty)^n} = C_0. \tag{5.6}$$

Since we are considering the number of real internal rate of return (IRRs) in the interval $(-1, +\infty)$ (real positive internal rate of return are located in the interval $(0, +\infty)$, the Descartes rule of signs applies to the interval $(-1, +\infty)$), the evaluation concentrates on Theorems 3, 4, and 5. The value of Theorem 3 relies on the number of significant changes in the cash flow stream. The value of Theorem 4 may be positive or negative. The value of Theorem 5 will always be $(+\infty, C_0 \prec 0)$. Thus, there are six potential solutions, as demonstrated in Table 5.2.

These results show us valuable information considering the number of real internal rate of return that may exist in the intervals $(-1, 0)$ and $(0, +\infty)$. As an example, let's evaluate the research case with an odd number of sign changes in the cash flow stream and $NPV(r = 0) \succ 0$.

In the interval $(-1, 0)$ Theorem 3 and Theorem 4 are above the horizontal axis. This means that the number of times that the *net percent value (r) function* crosses the horizontal axis shall be either 0, or an even number of times, as demonstrated in Figure 5.2.

Suggestion 1. If an investment project has an odd number of significant changes to its cash flow stream and $NPV(r = 0) \succ 0$.. If there is an even number of significant changes on its cash flow stream and $NPV(r = 0) \prec 0$ then there are either 0 or an even number of real IRRs in the interval $(-1, 0)$. It follows that an investment project with an even number of significant changes on its cash flow stream and $NPV(r = 0) \succ 0$. If there is an odd number of significant changes on its cash flow stream and $NPV(r = 0) \prec 0$ then there is an odd number of real IRRs in the interval $(-1, 0)$ (at least one).

Table 5.2 The potential value of crucial theorems on the interlude net present value function

Value of NPV(0)	Odd number of sign changes in the cash flow stream	Even number of sign changes in the cash flow stream
$NPV(0) \succ 0$	Theorem 3. $(-1 + \varepsilon, +\infty)$	Theorem 3. $(-1 + \varepsilon, -\infty)$
	Theorem 4. $(0, NPV(0) \succ 0)$	Theorem 4. $(0, NPV(0) \succ 0)$
$NPV(0) = 0$	Theorem 3. $(-1 + \varepsilon, +\infty)$	Theorem 3. $(-1 + \varepsilon, -\infty)$
	Theorem 4. $(0, 0)$	Theorem 4. $(0, 0)$
$NPV(0) \prec 0$	Theorem 3. $(-1 + \varepsilon, +\infty)$	Theorem 3. $(-1 + \varepsilon, -\infty)$
	Theorem 4. $(0, NPV(0) \prec 0)$	Theorem 4. $(0, NPV(0) \prec 0)$

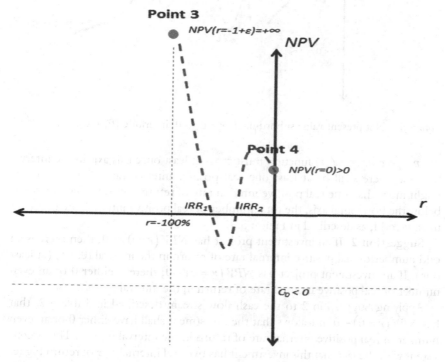

Figure 5.2 Net present value subordinate $(r = 0) > 0$, interlude $[-1, 0]$.

As an instance, consider the cash flow stream shown in Table 5.2. It has an even number of sign changes (2) *and net percent value* $(r = 0)$ is < 0 (it is –2). Based on Suggestion 1, this cash flow stream should have 0 or an even number of real internal rate of return in the interval $(-1, 0)$, which is consistent because it has 0 real internal rate of return on that interval.

Interval $(0, +\infty)$: taking a look at Theorem 4 (above the horizontal axis) and at Theorem 5 (below the horizontal axis) in Figure 5.3 it can be deduced that

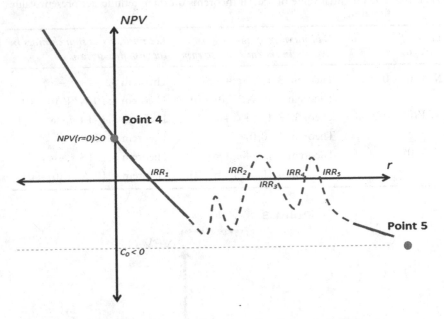

Figure 5.3 Net present value subordinate ($r = 0$) > 0, interlude $[0, +\infty)$.

the *net percent value* (r) function must cross at least once this axis in the interval $[0, +\infty)$. There shall be at least one real positive internal rate of return. There might more than one real positive internal rate of return, but since Theorem 5 is below the horizontal axis, the total number of real positive internal rate of return must be odd, as described in Figure 5.3.

Suggestion 2. If an investment project has $NPV(r = 0) \succ 0$, then there is an odd number of real positive internal rate of return in the interval $(0, +\infty)$ (at least one). If an investment project has $NPV(r = 0) \prec 0$, there is either 0 or an even number of real positive internal rate of return in the interval $(0, +\infty)$.

Applying Suggestion 2 to the cash flow stream described in Table 5.2, that has $NPV(r = 0) \prec 0$, it follows that the investment shall have either 0 or an even number of real positive internal rate of return in the interval $(0, +\infty)$. This is consistent with the fact that this investment has two real internal rate of return bigger than 0.

If $NPV(r = 0) = 0$ the number of real positive internal rate of return will depend on what occurs immediately to the right of Theorem 4. That is:

$$\lim_{r \to 0+\varepsilon} NPV(r) = C_0 + \frac{C_1}{1+\varepsilon} + \frac{C_2}{(1+\varepsilon)^2} + \ldots + \frac{C_n}{(1+\varepsilon)^n} \tag{5.7}$$

If this value is greater than 0, then there will be an odd number of real positive IRRs. If this value is smaller than 0, then there will be 0 or an even number of

Table 5.3 Depending on the value of net present value, the number of accurate positive internal rates of return

Number of real positive IRRs		
$NPV(0) \succ 0$	Odd number (at least one)	
$NPV(0) \prec 0$	0 or an even number	
$NPV(0) = 0$	If even the number of sign variations and $NPV'(0) \neq 0$	Odd number (at least one)
		0 or an even number
	If even the number of sign variations and $NPV'(0) \neq 0$	

Source: Self-elaboration.

real positive IRRs. Since it may be complicated to determine whether the limit reported in Equation 5.7 is bigger or smaller than 0, the following proposition is provided.

Suggestion 3. If an investment project has $NPV(r = 0) = 0$, an even number of sign variations, and does not report a maximum in $r = 0$ (that is, $NPV'(r = 0)$ $f = 0$), then the number of real positive internal rate of return is odd (at least one). It follows that if an investment project has $NPV(r = 0) = 0$, an odd number of sign variations, and does not report a minimum in $r = 0$ (that is, $NPV'(r = 0) f = 0$), then the number of real positive internal rate of return is either 0 or an even number. These three suggestions are summarized in Table 5.3.

5.6 Conclusion

This chapter demonstrates that when the sum of the cash flows of an investment project in nominal terms is greater than 0, it has an odd number of real positive internal rates of return equal to or smaller than the number of sign variations on its cash flow stream. This chapter also determines that the usual 3Ps investment project has a single sign variation on its cash flow stream, from negative to positive. This sign variation signals the end of the construction stage and the operation and maintenance phase.

Under these solutions, it can be argued that P3s have a single real positive real internal rate of return. The usual usage of the real internal rates of return as a criterion to assess the sustainability of 3Ps investments is thus validated. However, for 3Ps projects with singular cash flow structures - over more than one sign variant – the internal rates of return criterion should be used with caution in these circumstances. A further evaluation of the net percent value function, its coefficients, and its sign variations allows further suggestions regarding the potential number of real positive internal rates of return that a cash flow might present.

This chapter complements the existing context on 3Ps within the '3Ps policy' category, and in particular within the 'Risk management and fiscal analysis' category, as explained by Roechrich et al. (2014). Within the fiscal, academic context, this chapter reports a novel proposition that can be utilized to recognize if a cash flow has a single real positive root when other existing suggestions fail.

Under the pressure of the global economy, government investment in infrastructure public service construction becomes a vital method to stimulate the economy. Hence, investigating financing methods for these big construction projects, such as BOT,[4] PFI,[5] and 3Ps, is of great practical significance. This chapter analyzed 3Ps projects' option properties and set up the viewpoint that 3Ps can be recognized as a European call option. Later we built up an option pricing algorithm for 3Ps projects. We conclude that the project's value is more significant than that calculated by the old net percent algorithm through a case study analysis.

A future study could further discover the value creation in 3Ps projects from the opinion of the private sector and report a convincing reason why multiple real real internal rates of return for a specific cash flow are of practical usage. This new algorithm improved the operator of investment decisions and provided a new method to analyze 3Ps and other similar projects' investment decisions.

Notes

1 Rating agencies Standard & Poor's (S&P) and Fitch Ratings use the letters *"AAA"* to identify bonds with the highest credit quality, while Moody's uses the similar *"Aaa"* to signify a bond's top-tier credit rating.
2 Returns the probability of getting less than or equal to a particular value in a normal distribution (no cumulative).
3 **Sensitivity analysis** is a method for predicting the outcome of a decision if a situation turns out to be different compared to the key predictions. It helps in assessing the riskiness of a strategy. Helps in identifying how dependent the output is on a particular input value.
4 The **Black–Scholes model** is designed to value European **options**, i.e., **options** that cannot be exercised until the expiration day ... The derivation of the **Black–Scholes model** is based upon the assumption that exercising an **option** does not affect the value of the underlying asset.
5 The private finance initiative (**PFI**) is a procurement method which uses private-sector investment in order to deliver public-sector infrastructure and/or services according to a specification defined by the public sector.

6 Public–Private Partnership for infrastructure and combating corruption

6.1 Transportation Infrastructure and 3Ps Efficiency

The choice of investment approach has a great effect on the performance of transport infrastructure. Positive projects like the 'Subway plus Property' model in Hong Kong have generated sustainable fiscal benefits for public transport projects. Owing to public debt and other constraints, 3Ps was proposed as an innovative investment model to address this problem and assist in developing transport infrastructure. Few studies provide a deeper understanding of 3Ps approach connections and transport projects (particularly the whole transport system). The research scope of this chapter is the regional network of the highway. With a famous 3Ps algorithm, travel demand prediction method, and relevant parameters as input, agents in a simulation framework can simulate the 3Ps highway choice over time. The simulation framework can analyze the relationship between the 3Ps approach and the regional highway network's performance. This chapter uses the Highway Network of Yangtze River Delta (FN-YRD) in East Asia as the context. The results display the value of using simulation algorithms of complex transportation systems to assist decision-makers in choosing the right 3Ps projects. Such a tool is essential given the ongoing transformation of the Chinese transportation sector's functions, including franchise rights of transport projects and the highway charging mechanism.

A Public–Private Partnership is a contractual scheme under which the public party and private firms cooperate and share risks and benefits to construct infrastructure projects or provide public products and services. According to the potential contribution to decreased transaction costs, innovation, continuous exploitation of a learning curve, the re-concentrating of government on its core tasks, and the enabling of large infrastructure investments, 3Ps has been widely applied in projects of transport infrastructure like rails, airports, roads, seaports, waterways, etc. (Cruz and Marques, 2013a; Siemiatycki, 2011). Additionally, the recent liberalization in the transport sector and the global economic crisis favor implementing transport projects based on the 3Ps algorithm (Tsamboulas et al., 2013). There are many successful cases, like the 'Subway plus Property' model and the 'Landlord Port' model.

DOI: 10.4324/9781003177258-6

Different models have been improved to implement 3Ps projects in the field of transportation engineering. According to the involvement of the private sector and risk allocation between the public and private party, the models can be classified into 12 types and grouped further into 4 categories: operations and maintenance, concession (public ownership of the facilities), concession (private ownership of the facilities), and full privatization (Percoco, 2014). The private party can get involved in a transport 3Ps project at different phases, like the beginning of design, construction, financing, operation, or maintenance, even through the whole project life cycle. Some 3Ps algorithms, like build-operate-transfer (BOT) and build-own-operate (BOO), focus on the transport projects' construction quality. The Sines Container Terminal in Portugal and the Valencia Cruise Terminal in Spain were constructed under this 3Ps algorithm (Roumboutsos et al., 2013). Some models, such as operating concessions, tend to involve the private sector during the operation phase to improve service quality. Many 3Ps projects in public urban transportation like the Line 4 Subway project in Beijing represent this sort (De Jong et al., 2010). Some models such as full/partial privatization or design-build-finance-operate (DBFO) are adopted due to the lack of public finance to provide transport service in an earlier stage. The M6 Tollway in the UK implemented using the DBFO is one instance of this sort (US Department of Transportation, 2007). Therefore, various 3Ps algorithms have been extensively applied in the construction of transport infrastructure. But the approach, which infrastructure should accept the 3Ps algorithm, remains an unresolved problem.

If the 3Ps approach were made for a transportation system, the users, operators, planners, and owners would constitute a set of distinct stakeholders, with each stakeholder making strategic decisions and investments to fulfill its objectives for system performance. Ultimately, however, transportation system performance is a function of the interactions between and the decisions taken by all stakeholders. These interactions can also complicate efforts at solving the choice problem of the 3Ps highway. This chapter approves the strategy of agent simulation required for the performance of transport infrastructure. The simulated performance can then analyze the validity of the 3Ps approach. The Highway FN-YRD in East Asia was selected as the simulation object. At this moment, the 3Ps approach can be applied to improve any roads, bridges, or tunnels. Major reasons for the choice of FN-YRD include: (1) the region of FRD is an advanced area in East Asia. It has achieved the level of middle-developed countries in terms of GDP and road network density per capita. Therefore, its results are useful in improved countries. (2) The FN-YRD has been improved after East Asia's 'reform and opening' policy by the end of the 1980s. But the regional prosperity was achieved under a political system that is not yet sound. This experience might be valuable for the countries whose political system improvement is similar.

The early attempts at using the 3Ps algorithm to build up transport projects were found in the late 1970s via highway concessions in France and the mid-to-late 1980s in Spain and England. The most vital impetus fostering transport 3Ps projects occurred in the 1990s in the UK, where economic reforms encouraged several efforts to privatize major elements of the national transport systems. Under the PFI, legislative and regulatory reforms were put into place to carry 3Ps projects primarily concentrated on the transport infrastructure, including railroads, public transportation, and aviation (US Department of Transportation, 2007). Since then, the 3Ps usage has spread quickly worldwide, first into other developed countries such as many European countries, the United States, Australia, Canada, etc., and later into developing ones in Asia, South America, and other regions.

Along with the worldwide adoption of the 3Ps shape to improve transport infrastructure, an increasing number of studies and reports have been published. By reviewing the literature, different focuses are found among these scholars. Some made efforts at summarizing the critical success elements of 3Ps usage in common (Thomas Ng et al., 2012; Mu et al., 2010; Yun et al., 2015) or the effects of certain factors like the institutional element (Panayides et al., 2015; Percoco, 2014; Verhoest et al., 2015). Some literature focuses on specific parts of the transport sector such as airports (Farrell and Vanelslander, 2015), ports (Cabrera et al., 2015; Macario, 2014), construction (Tang et al., 2010), or urban transport (Willoughby, 2013). Other study directions include 3Ps contract and negotiation (Cruz and Marques, 2013b; Domingues and Zlatkovic, 2015; Hart, 2003; Krüger, 2012; Xu, 2010), risk allocation, and assessment or mitigation (Chan et al., 2011; Li et al., 2005; Vassallo, 2006). Beyond that, many publications are focusing on discussing the performance of transport 3Ps projects. Compared to the traditional financing styles, 3Ps projects benefit from performing transport services on time and on budget, gaining efficiency and effectiveness, and reducing overall construction and operation (Cruz and Marques, 2013a; Grimsey and Lewis, 2002). The 3Ps algorithm's overall success in terms of time, cost, and quality for stakeholders (public, private, and user) was involved by analyzing four 3Ps transport projects from four different EU countries using the strategy of qualitative comparative analysis (Liyanage and Villalba-Romero, 2015). Service quality is even ranked as the most critical factor when the government chooses the 3Ps algorithm (Tsamboulas et al., 2013). The UK Treasury estimates that using the 3Ps algorithm can produce a cost-saving of 17–25% on average across all sectors (Alfen et al., 2009). Similar results are also provided by evidence from Australia. The 3Ps algorithm has cost-saving benefits of 9–23% and on-time delivery over old ones (Infrastructure Partnerships Australia, 2007). Transport infrastructure needs high investment and will increase the burden of public deficit. The 3Ps algorithm provides an alternative through the private sector's involvement and delivers the transport service faster by avoiding inflationary cost increases (US Department of Transportation, 2007).

Furthermore, the 3Ps algorithm fosters innovation. It provides a flexible way to charge transport service tolls. Besides the traditional mileage, other criteria, such as vehicle types in terms of emission volume or size, occupancy level, and travel time (peak time vs. off-peak time), can increase transport service usage and avoid congestion and pollution (Tamayo et al., 2014).

Overall, most previous and current publications on the effect of the 3Ps algorithm on transport infrastructure performance mainly focus on the construction phase. Few are found to evaluate the performance after construction.

Further, public investment decisions tend to be made in the short term. A feasibility study of a 3Ps project is generally conducted when a transport project is to be initiated. A single transport project has a limited economic benefit. The relevant evaluations are not comprehensive, and the rationality of decision-making is sensitive. Some transport projects that are feasible during the analysis phase finally prove to be failures after implementation. For example, the 3Ps project of Hangzhouwan Bridge was a great success at the beginning after construction in 2008 as it decreases the distance between Ningbo and Shanghai by 30%. Unfortunately, the *high-speed railway* (HIGH-SPEED RAIL) was put into practice, and the Jiashao Bridge (a neighboring bridge) was also built in 2013. Massive numbers of travelers have been attracted by the Hangzhouwan Bridge. As an outcome, many private investors have had to leave the 3Ps project of Hangzhouwan Bridge. The share belonging to the private party reduced from above 50% in 2009 to approximately 15% in 2013.

Therefore, governments must have a comprehensive analysis of the effect of scale on a series of transport 3Ps projects to achieve reliable decisions. However, it is quite complicated to analyze the whole regional highway network. When examining the network's performance under the 3Ps approach, many factors, including the highway feature, travel modes, travel behaviors, and other travel modes' analysis mechanisms, shall be considered. Hence, it is significant to establish the analysis algorithm of the performance of transport infrastructure.

This chapter simulates the evolution of performance indexes for transport infrastructure. Agent-associated modeling methodology has a long lineage, beginning with von Neumann's (1966) work on self-reproducing automata. Agents are 'objects with attitudes' (Bradsha, 1997). The application of agents in the transportation field is widespread, e.g., traffic control using agent simulation (De Oliveira and Camponogara, 2010). However, a few research types predict transport infrastructure's performance by comprehensively considering the traffic system's complex interaction. We will conduct this kind of research. Although many scholars have focused on transport infrastructure performance, most do qualitative analysis and lack the model structure. This kind of study does not show a transplantable character.

This section's innovative contribution is as follows: the interaction relationship between the 3Ps approach and the settled objects is considered when doing the feasibility study. That is an improvement compared to the traditional algorithm (conducting single project evaluation only). The subsequent portions are organized as follows. A dissection analysis of the impact elements of the 3Ps algorithm on the transport infrastructure is first conducted. According to the analysis, an agent-based simulation framework is established to make a 3Ps approach. The simulation results are then reported and discussed.

6.2 Urban Infrastructure and efficiency of 3Ps

The decision-maker should understand the impact of the 3Ps approach on the whole transportation system to make a better decision about each road's investment model. Generally, transportation departments need to schedule the highway investment each year. The 3Ps approach should also be determined year by year. In our model, the 3Ps approach shall be determined for each year's roads. As denoted in Figure 6.1, nth year's roads mean n is the start year of the constructed roads. If the 3Ps algorithm invests in a specific road, its construction and service feature would be affected. This effect can propagate to the whole network. The network structure and road impedance of the following years would be varied. Changes also happen in travel demand. Finally, the transportation network performances, containing the assigned traffic flow and economic indexes, would be varied.

Specifically, the 3Ps algorithm can impact the corresponding road's performance in four ways: construction time, facility quality, service price, and service level (see Figure 6.1).

- **Construction time**

Evidence from actual transport 3Ps projects demonstrates that the 3Ps algorithm has the benefit of decreasing the construction time compared to the traditional investment model. The infrastructure facilities can be thereby put into operation earlier to meet the travel demand. The values of the construction period presented in Figure 6.1 represent the possibility of accomplishing the 3Ps project. For example, setting $x = 75$ means that the 3Ps algorithm can reduce the construction period to 75% of the traditional investment model.

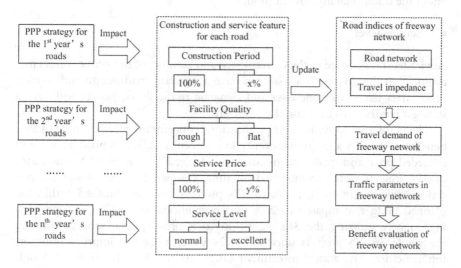

Figure 6.1 The protocol for assessing the impact of the 3Ps approach on the highway.

- **Facility quality**

One benefit of the 3Ps algorithm is that both public and private sectors attach great importance to transport projects' overall benefits. Therefore, enough funding will be arranged for the whole project life cycle, based on construction to maintenance. In other words, we generally find some phenomena under the traditional investment model, such as project funds' corruption. That will lead to the waste of fiscal and human sources, consequently affecting the construction quality. Road quality is one crucial element affecting a user's decision. Given two highways in parallel, travelers will undoubtedly select the one with better conditions than those full of potholes when other conditions are the same.

- **Service cost**

Price is the direct element influencing people's travel demand. In general, the government has a fixed and unified charge mechanism towards the highway, which neither increases the travel demand nor improves operation profits. The highways in some developing countries such as East Asia don't have the same problems as in the United States – traffic congestion. The highways in major regions in East Asia have a massive capacity to serve potential travel demand. More price-sensitive passengers may be attracted by setting up flexible charging fees. Besides, the high-speed rail has a comparative advantage over a long distance. If the highway doesn't have any effective charging fees, it will undoubtedly lose market share in long-distance travel. The value of service cost presented in Figure 6.1 represents the possibility of charging users. For example, setting $y = 90$ means that the 3Ps algorithm can decrease the charging fee to 90% of the one under the traditional investment model.

- **Service level**

Service level is denoted as the convenience and comfort of the whole travel environment to the drivers. If a road can have a standard road design and provide timely data like work-zone data and dynamic traffic flows, people will have a stronger tendency to travel on this highway.

There are a good number of publications concentrating on the quantitative benefits of the 3Ps algorithm. Seventy-five percent of UK 3Ps projects met and exceeded price and quality requirements and saved 17% in costs. Further, 80% of 3Ps projects were accomplished on time, compared to 30% under the traditional investment model; 80% of 3Ps projects could be finished within the planned budget, compared to 25% under traditional ones. Chile is one leading country utilizing the 3Ps algorithm to improve public services. Among the whole 36 3Ps projects since 1994, 24 were utilized to improve transport infrastructure. The annual investment ranges from 0.3 to 1.7 billion US dollars. By screening international publications in the transport 3Ps projects sector (Alfen et al., 2009; Infrastructure Partnerships Australia, 2007; Liyanage and Villalba-Romero, 2015; Tamayo et al., 2014; Tsamboulas et al., 2013; US Department of Transportation, 2007), it is found that the following features of the 3Ps algorithm are popular: construction time (75%), facility quality (flat), service price (90%), service level (excellent). To test the 3Ps approach's validity in FN-YRD, the 3Ps algorithm parameters are taken as input for the simulation algorithm. The traditional investment model is utilized as a reference. Its features are as follows: construction time (100%), facility quality (rough), service price (100%), service level (normal).

6.3 Simulation algorithm

The simulation algorithm is conceptually an agent-associated model with four components: an environment, rules, agents, and outputs. The simulation comprises three modules (the travel demand prediction module, the project evaluation module, and the 3Ps investment strategy module), illustrated in Figure 6.2 and described below. The environment is reflected in the travel demand prediction module. It is reported by each city's population in YRD, the FN structure under different investment strategies, and travel demand. Rules are embedded in the project analysis module and the 3Ps investment approach module to calculate indexes and implement decisions. Agents include government, private enterprise, and road users. Outputs include a set of results of the investment decision of the 3Ps algorithm on the FN.

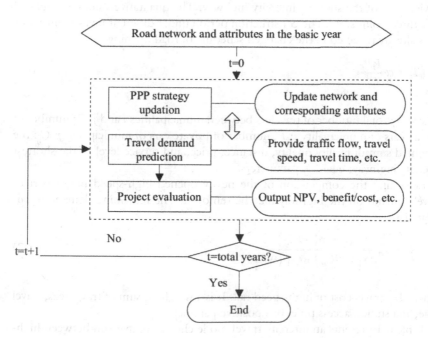

Figure 6.2 Structure of a model.

The highway has a history of over 30 years in East Asia. However, the high-speed rail has been open only since 2004. To investigate the development of the highway under the influence of high-speed rail, ten years (2005–2014) were chosen based on the availability of data. Each run through the modules above represents one year. The strategy module is used to update the investment model for the constructed roads in the studied year. The varied strategies would correspond to the different road networks and road attributes in the future. We can implement the travel demand prediction and project evaluation modules under the assumed 3Ps approach to get the required network performance. This performance can be utilized to evaluate and update the current 3Ps approach this year, conversely. This inner loop procedure would never stop until the network performance can satisfy the 3Ps approach requirement. At the outer loop, the year would increase gradually. The following sections discuss each module in detail.

6.4 Demand to Travel and divination Algorithm

This module predicts the traffic flow volume on each link in FN-YRD for each year in the simulation. This procedure is accomplished with a simplified aggregate four-step model, with travel generation and distribution according to a formula's traditional gravity function (Equation 6.1). The numerator variables represent the number of private cars, as car owners generate the most travel demand. The denominator is reported via a weighted sum of distance, facility quality, price, and service level of the shortest intercity highway. The qualitative values are defined on a three-point scale. The construction period under different investment models is already reflected in the changed schedule of road opening.

$$T_{ij} = \alpha \frac{P_i P_j}{C_{ij}^b} \tag{6.1}$$

Where T_{ij}: total number of travelers between municipalities i and j; P_i: number of private cars of municipality i; P_j: number of private cars of municipality j; C_{ij}: the weighted sum of facility quality, distance, price, and service level of the shortest intercity highway; a and b: parameters.

Regarding the competition of the newly opened high-speed rail, a part of travel demand on the highway may be removed. The remaining demand is calculated as:

$$R_{ij} = T_{ij} \frac{\exp(-\theta C_{ij})}{\exp(-\theta C_{ij}) + \exp(-\theta D_{ij})} \tag{6.2}$$

Where D_{ij}: travel cost of high-speed rail. It is a weighted sum of travel fees, travel time, and station access time. θ: a positive parameter.

Table 6.1 presents an intercity travel mode choice comparison between high-speed rail and the highway. The results are created according to the mathematical

Table 6.1 Turbo train or highway as an intercity mode of Urban Infrastructure

Factor	Hangzhou–Shaoxing	Ningbo–Shaoxing	Nanjing–Huzhou
Distance (km)	40	110	220
Travel fees for HIGH-SPEED RAIL (¥)	19.5	51.5	85
Travel time by HIGH-SPEED RAIL (m)	20	40	60
Station access time (m)	40	30	40
The ratio choosing the highway based on simulation (%)	89	73	67
The ratio choosing the highway based on the investigation (%)	92	72	61

Notes: the fee to use the highway, including toll and fuel, is generally ¥ 1/km; 'm' stands for minute.

algorithm and investigation separately. A non-significant difference demonstrates the reliability of the mathematical algorithm.

As demonstrated earlier, all trips are assigned to the FN, an assignment that reflects the dominance of the auto mode for intercity travel in YRD. Finally, trips are assigned to the path via the utilization of an incremental assignment strategy. This algorithm gets a result approximate to that of the equilibrium traffic assignment. It follows the principle that the traveler's priority route is the shortest highway. Only if it is capacity constrained is the second shortest route is under consideration.

6.5 Analyzing the 3Ps module

NPV is utilized to evaluate the effectiveness of the 3Ps approach. Net percent value is defined as the difference between the present values of incoming and outgoing cash flows over some time. This algorithm has taken the operating profits and the internal rate of investment return into consideration; thus, it complies with financial effectiveness evaluation. Therefore, it can use the net percent value calculation formula (Equation 6.3) to evaluate the 3Ps approach.

$$NPV = \sum_{i=1}^{n} \left[\left(CI_p - CO_p \right) \left(1 + m \right)^{-1} \right] \qquad (6.3)$$

Where n: a total time of the investment style; CI_p, CO_p: denoted incoming and outgoing cash flows of each year of operation sectors; m: discount rate.

6.6 3Ps Algorithm

Mathematically, the 3Ps approach is a selection problem with 2^ρ possible combinations. The variable ρ herein represents the number of planned roads. Just 30 roads would increase the calculation counts of travel demand prediction and project evaluation by billions. Therefore, we design a heuristic method to solve this issue rapidly. The 3Ps approach is solved year by year in our method. The year-by-year method is practical because we do not know the road planning of the future. For instance, if we need to determine a 3Ps approach for the constructed road this year, we cannot consider the impact of roads which may be planned in the future. Besides, we make an assumption when counting net percent value. Except for the current and previous years, the future years' planned roads are assumed not to be considered. Based on the calculated net percent value, the current year's 3Ps approach could be updated with several rules. Then, we would come back to the step of the net percent value calculation. This iteration procedure continues until the updated 3Ps approach in the current year can meet the requirement. Our case application would validate this method. The following sections would be utilized to set the 3Ps approach of the current year.

  3Ps investment decisions are the results of interactions of multiple agents. The government decides the constraint of the public budget and chasing the profit of highway projects. Private enterprises decide based on the internal rate of return. Users' requirements are based on the total travel cost. To some extent, both government and private enterprises prefer highway projects with higher profit rates than those featuring negative net percent value. In terms of projects with negative NPV, the public welfare function forces government to undertake the construction accountability of highway projects with negative net percent value. In terms of positive net percent value, the government usually tends to operate those with higher profits and transfer those with lower profits because the latter may risk operating at a loss. Consumers expect private enterprises to operate all the highway services with high operation quality in the user's view. Considering all three agents' interests, the private enterprises should take the low-profit projects, and the government needs to transfer sections of high-profit projects to private enterprises. This principle is utilized for the 3Ps approach to the current year's projects A:

Step 1: each project within A is regarding the investigation into the 3Ps approach.
Step 2: predict the incoming and outgoing cash in the operation years; gain the net percent value for projects within A; if Step 1 is not the first step, turn to Step 4.
Step 3: projects with positive net percent value are chosen each year and ranked from low to high as alternatives for the 3Ps investment decisions; the first 80% of all the alternative projects each year will be selected for the 3Ps algorithm; the remaining projects are for traditional investment; turn to Step 2.
Step 4: if each 3Ps project is profitable, end this year's simulation and turn to next year's simulation; otherwise, transfer the projects without benefit to traditional investment; turn to Step 2.

Table 6.2 Simulation algorithm variables

Module	Variables	Values
General	Start year	2005
	End year	2014
FN	Design speed: low	100 km/h
	Design speed: middle	110 km/h
	Design speed: high	120 km/h
	Traffic capacity: low-speed	2000 pcphpl
	Traffic capacity: mid-speed	2200 pcphpl
	Traffic capacity: high-speed	2400 pcphpl
Travel demand prediction	a	0.005
	b	1.6–2.8 (depending on the sample data)
	θ	0.5
Project evaluation	Evaluation time horizon	10 years
	Discount rate	5%
	Traveler value of time	¥ 50/h
	Peak-hour traffic in the peak direction	65%
	The ratio of traffic in peak hour	13.5%

6.7 Project data

YRD is taken as the simulation object. The input data are presented in Table 6.2. The start and end years are 2005 and 2014, respectively. For the design speed, most of East Asia's highway is limited to 100–120 km/h. Three intervals of design speed are set in this case. When vehicle density is the same, traffic capacity will correspond to the design speed. Referring to the traffic engineering manual, the corresponding traffic capacities are listed in Table 6.2. The parameters of travel demand prediction refer to the traffic planning experience in Chinese cities. The parameters in project analysis are described as follows. The discount rate reflects the time value of cash. The value of time in traveler's opinion can be defined as how much does time merit for them. The survey investigates it. The peak-hour traffic in peak direction reflects the asymmetrical distribution of traffic flow on the road. The ratio of traffic in peak hours reflects the amount of traffic flows in peak hours. By the end of each year (each simulation run), the attributes such as the FN and OD matrix will be upgraded for the next step simulation. This process continues until the end of the tenth year. Relevant data will then be illustrated for more in-depth analysis.

6.8 Project Results

This section gives an opening time comparison of the FN in YRD under the traditional investment strategy and the 3Ps approach. The start year for the construction of each road is attached in the middle of the figure. Roads labeled in bold and italic are selected for the 3Ps algorithm. These roads are sketched with dotted lines in the right figure.

Table 6.3 Metrics for the 3Ps approach and the conventional strategy

Performance metric	Results of 3Ps approach	Results of traditional strategy
Number of trips per day (2014)	1,470,000	1,260,000
Average travel distance (2014)	195 km	156 km
Average peak travel time (2014)	1.86 h	1.61 h
Average peak speed (2014)	105 km/h	97 km/h
Average daily total delay (2014)	410,000 h	467,000 h
Total investment (2005–2014)	268 bil. CNY	224 bil. CNY
NET PERCENT VALUE	109.9 bil. CNY	87.3 bil. CNY

The output data are summarized in Table 6.3 and the performances of the 3Ps approach and the traditional one are compared. The 3Ps approach has a comparative advantage in most performance indexes, indicating the feasibility of adopting the 3Ps algorithm in transport projects. An exception is the average trip period. The trip period is longer under the 3Ps approach because the travel distance is increased. It also indicates that travel convenience is improved under the 3Ps approach; thus, people are prone to increase their travel distance. net percent value in the table is a significant analysis metric and analyzed in detail in the following sections.

6.9 Analysis of demand change under the 3Ps approach

This section indicates that the travel demand change between typical counties connected by both high-speed rail and highway. Overall, travel demand has increased remarkably. The reason is that car ownership in YRD has experienced a time of high growth since 2005. Furthermore, the years of 2007 and 2013 catch particular attention in travel demand improvement because the high-speed railway was put into operation, which attracted a section of previous vehicle travelers.

There is a temporary demand decline between Shanghai and Suzhou when the high-speed railway between Shanghai and Nanjing[1] was opened in 2007. However, the travel demand soon increased steadily as the highway has a comparative advantage over the railway over short travel distances (80 km). Besides, the rapid increase in trip demand could also be attributed to the auto auction. The auction price for license plates boomed recently in Shanghai, making many Shanghai workers choose to buy cars and live in Suzhou. Subsequently, large traffic volumes are designed between these two cities.

The trip demand between Ningbo and Jiaxin was not great a few years ago. The cargo traffic accessing Ningbo Port played a significant role. The Hangzhouwan Bridge opened in 2008, developing traffic accessibility, and enhancing the travel demand, as it shortens the trip distance from 180 to 120 km. But the Jiashao Bridge, a neighboring bridge crossing Hangzhou Bay, opened in 2013, has a significant impact on Hangzhouwan Bridge's strengths in cost and time. In addition to that, a high-speed railway was operated between these two cities in 2013.

Both have resulted in the decline of car traffic rather than an increase. This demonstrates that the risk of recouping highway construction costs increases when high-speed rail appears. Only those highways (e.g., Hangyong Highway) operated for an extended period could recoup the investment. Nowadays, alternative travel modes bring fierce competition to the trip demand of highway, to impact its profits. Therefore, it is essential to consider these effects when making investment decisions.

Before 2008, people had to drive 300 km of the highway to get from Nanjing to Huzhou. Although there is a direct provincial highway connecting the two cities via a shorter distance, few travelers select this option because of the many signal-controlled intersections. It would increase travel time and the risk of a traffic accident. Luckily, the direct highway was opened in 2008. It decreases the distance to 200 km and increases the travel demand significantly. Although the HIGH-SPEED RAIL opened in 2013 has affected vehicle travel's growth rate, its demand is increasing steadily, which differs from the Hangzhouwan Bridge. One possible reason is that the Tai Lake between Nanjing and Huzhou makes it impossible to choose other bypassing lines.

6.10 Analyzing the investment profit

Following the step of counting net percent value for each highway, the profit/cost on each highway of different investment models is reported in Figure 6.3. Most highways have a higher profit/cost ratio under 3Ps investment. In addition to the traditional strengths of the 3Ps approach, one external reason plays a critical role. The 3Ps approach shortens the construction period of highways and thus brings forward the date for the operation. This advantage reduces external factors such as the high-speed rail because it may be a race against time to recoup highway construction costs.

This section defines the research scope of a regional highway network. The impacts of the 3Ps approach on the performance of transport infrastructure are

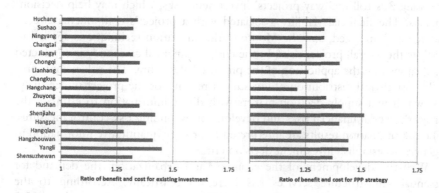

Figure 6.3 Value of benefit/cost on each highway.

analyzed. The agent-based simulation approach is introduced to examine the relationship between the 3Ps approach and a highway network's performance. Further, this chapter compares the results with the traditional investment strategy by FN-YRD in East Asia. Some conclusions are achieved. For instance, the 3Ps approach can indeed increase the benefit of the whole highway network. The conclusions can be used for applied research in transport 3P projects. Besides, our future focuses are as follows.

[1] The prediction accuracy of the traditional gravity four-step model may not be high. Given sufficiently detailed data in the future, the activity-based prediction method could be borrowed to achieve accurate travel demand.
[2] The algorithm of net percent value utilized to analyze the 3Ps approach doesn't consider the impact of risks. This must be improved in future research.
[3] The method of choosing the highway to use the 3Ps algorithm is practical but also rough. To make more scientific and logical decisions, it is important to use the Eq theory to create a general formula in future work.

6.11 A simulation approach to Urban Infrastructure

The demands for delivering highway services continue to grow worldwide. However, funding from government and public agencies alone cannot provide the capital required to operate and maintain existing highway systems, much less to construct new ones. 3Ps are an innovative funding mechanism for highway agencies to utilize private capital and expertise in transportation infrastructure projects to increase funding options to bridge the budget gap. Even based on sectors involved in 3Ps take different paths and responsibilities, there are still risks taken or shared with the public and private parties. Specifically, assessing risks associated with investments' potential returns is of great importance to the private and public sectors. This section presents a methodological framework for assessing 3Ps toll highway projects' investment risks, which may help decision-makers. The financial viability associated with a project's components is considered and analyzed, and the Monte Carlo simulation technique is applied to analyze the overall project risks. In the end, a numerical case study is conducted to demonstrate the application of the proposed algorithm. The risk analysis provides a statistical distribution of investment returns for the project under analysis, which will supply decision-makers with direct information to estimate the project's overall financial risks and develop corresponding risk control measures. The risk simulation results are interpreted to provide quantitative information to agencies to establish investment decision criteria.

With the development of the economy and urbanization, the demand for essential infrastructure services has increased drastically. According to the American Society of Civil Engineers (ASCE), it was estimated that about $2.2 trillion would be needed in the next five years to improve US infrastructure from its current average score of '*D*' (ASCE, 2009). Besides, travel demand and high expectations from the public concerning mobility and service level will continue to increase (Brown, 2007; Ortiz et al., 2008). However, public funds alone, which usually come from funds explicitly designated for infrastructure or general government funds, cannot cover all infrastructure construction, operation, maintenance, and replacement costs. There is a call for the use and involvement of other capital sources. Under these circumstances, 3Ps, which introduce private capital and expertise into a project, have been applied to address this budgetary shortage problem (Pagano and Perry, 2008; Huang, 2016; Han, Z., 2013, 2017). Although 3Ps are seen as an innovative funding mechanism, traditional public financing and procurement methods have slowly but steadily given way to private involvement in delivering public infrastructure services.

Three parties are always involved in a 3Ps project: the public authority, the private investor, and the lender (e.g., banks). These parties work together by undertaking or sharing different tasks. There are different types of 3Ps according to different classification criteria. According to how the raised debt is repaid, 3Ps can be categorized as private finance initiatives or concession contracts (Yescombe, 2011; Pantelias, 2009). The 3Ps can also be seen as availability-based or usage-based according to the distribution of risks among the private sector and the public authority.

Furthermore, according to the legal position of the private sector involved in the project, 3Ps can be classified as build-own-operate-transfer, build-operate-transfer, build-transfer, and joint ventures (Yescombe, 2011; Grimsey and Lewis, 2007). 3Ps can involve existing brownfield projects (i.e., the lease of an existing facility), or they can involve proposed new facilities, which are known as greenfield projects. In the case of a brownfield project, a public entity generates a capital inflow or debt payoff by transferring the rights, responsibilities, and revenues attached to an existing asset to a private-sector entity for a defined period. For greenfield projects, a public agency transfers all or part of the responsibility for project development, construction, and operation to a private-sector entity.

Because of the ability to attract private investments and cost-effectiveness, 3Ps are becoming popular worldwide. They have been applied in various fields, such as wastewater treatment, environmental infrastructure, power plant construction, transportation, and telecoms, as well as many other areas (Wang et al., 2000; Sachs et al., 2007). By 2011, in the United States, 24 states, including Texas, Florida, and California, and the District of Columbia had used the 3Ps process to help finance and construct at least 96 transportation projects worth $54.3 billion (Reinhardt, 2011).

Of those implemented projects, toll highway infrastructure projects are an essential part. The first major toll 3Ps project in the United States is the E-470 tollway project, located east of the Denver–Aurora, Colorado, metropolitan area; construction began in July 1989 through a $323 million design-build contract. User fee collection was the primary revenue source for that project and is expected to meet the required revenue. After that project, more 3Ps toll highway projects were signed and constructed, including the Chicago Skyway in 2005, the

Indiana toll road, the Virginia Pocahontas Parkway in 2006, and the Colorado Northwest Parkway in 2007. Several highway 3Ps projects have been approved since 2008, accounting for about 11% of the total investment in new highway capacity provided by the national capital in 2011. Most of the projects are express lanes that can be tolled and are constructed next to existing highways in heavily congested urban areas.

Although 3Ps have been proved as an effective way to provide funding flexibility and relieve budget shortfalls, not all 3Ps are successful experiences. Failure cases have occurred during the development and exploration processes because the investment risks were not adequately estimated. For example, the M1/M15 motorway, Hungary, built in 1995, was the first toll 3Ps motorway tendered and implemented in Central and Eastern Europe. However, after it was opened to the public, the traffic volumes were about 40% lower than anticipated. As a result, the concessionaire was unable to service its debt, and ultimately the government had to take over the concession at a high cost (Cuttaree, 2008). The South Bay Expressway, SR-125, is a 13-mi 3Ps toll road east of San Diego, California, that runs north–south, beginning near the Mexican border. As a result of the housing crisis, recession, and slowdown of truck traffic from Mexico, traffic and revenue fell more than 50% short of projections.

Furthermore, the project experienced a series of problems that increased costs, including that the construction took 41 months longer than anticipated. Consequently, the San Diego Association of Governments had to purchase South Bay Expressway back from the private operator (Minnesota Department of Transportation, 2016). Therefore, it is essential to thoroughly understand the risks associated with 3Ps projects and develop quantitative methods to fully assess those risks.

6.11.1 *3Ps and Risks*

Risks exist in different phases during a project's life cycle. Compared with traditional publicly financed projects, the government and private-sector providers' risk allocation is much more complicated in a 3Ps project. Both parties should clearly understand the various risks involved and agree to an allocation of risks between them. According to Edwards and Bowen, the notion of risk is a human construct that refers to the probability of the occurrence of a special adverse event during a stated period together with the quantification of its consequences (Edwards and Bowen, 2003). The risks can be categorized according to the source or origin and time of occurrence during different project phases. Traditional classification divides risk into external and internal based on the origin of the risk factors. Internal risks are risks within the project and are affected by decisions made concerning the project; examples include decisions on construction risk and maintenance risk. External risks come from outside the project and usually are hard to control; examples include a sudden change in the economic situation and political issues (Songer et al., 1997).

The United Nations Industrial Development Organization distinguishes the risks into general and country risks and specific project risks. General and country risks include political, commercial, and legal risks; specific project risks include developmental, construction (completion), and operating risks (Kalidindi and Thomas, 2003, p. 317; Jeon and Amekudzi, 2007). A more generally accepted classification, life cycle risk, is based on the project phases that the risks belong to. These risks fall into four categories: development, construction, operation, and ongoing (Songer et al., 1997). The risks can also be classified as systematic and nonsystematic and specific and nonspecific, as well as government-, sponsor-, lender-, contractor-, and user-related risks (Asenova and Beck, 2003; Xenidis and Angelides, 2005). In current industry practice, risk classification is usually more project-related and involves combining the above methods. Generally speaking, greenfield projects present higher risks than brownfield projects because of more significant uncertainties surrounding traffic forecasts, permitting, and construction.

This section focuses on the assessment of the investment risk of 3Ps projects. Investment risk is a financial-type risk, which is defined as the probability of failure to secure the required infrastructure-generated revenue used for servicing debt (as a minimum requirement), or failure to obtain an adequate return on the investment, or both (Kakimoto and Seneviratne, 2000). Failure to meet either of the two goals above can be deemed as a financial project failure. The investment risk directly depends on infrastructure-generated costs and revenues. It is critical in determining a 3Ps project's overall financial viability based on the expected operation characteristics and the proposed financing scenarios. Assessing the investment risk correctly and precisely to ensure the successful procurement of 3Ps projects is critically important.

Various studies have been carried out to develop methods to assess and quantify the financial viability of a project, such as the NPV, the (IRRs), the benefit-cost ratio (profitability index), the return on equity, and the payback period. Of these methods, the net percent value and internal rate of return are the most popular and widely accepted criteria used to measure the profitability of capital investments, with the profitability index applied in a secondary level of analysis (Keown et al., 2005). net percent value is the difference between the current value of the expected cash inflows and the cash outflows, which compares the value of a dollar today with the value of that same dollar in the future, taking inflation and returns into account. internal rate of return gives the profitability of an investment when net percent value equals zero. net percent value solves the present value of a stream of cash flows, given a discount rate, whereas internal rate of return solves for a rate of return when the net percent value is set equal to zero. Thus, net percent value and internal rate of return are related and will move together. A negative net percent value indicates that the IRR is less than the cost of capital, and the investment should be rejected; a positive net percent value implies that the internal rate of return is more than the cost of capital, and the investment should be considered.

Several researchers used real options models to valuate infrastructure investments and 3Ps projects (Lee, 2011; Blank et al., 2009; Rakic' and Rad-enovic,' 2014; Vajdic' and Damnjanovic,' 2011; Vandoros and Pantouvakis, 2006, p. 594). Ye and Tiong proposed an net percent value at-risk method by combining the weighted average cost of capital and dual risk-return methods (Ye and Tiong, 2000). Mohamed and McCowan developed a method using interval mathematics and possibility theory to model the effects of monetary and nonmonetary aspects of an investment option (Mohamed and McCowan, 2001). Jafarizadeh and Khorshid-Doust suggested the framework of mean–semi deviation behavior, which has the advantage in the collective evaluation of the firm's risk by all market participants (Jafarizadeh and Khorshid-Doust, 2008). Pantelis and Zhang evaluated revenue-generating transportation infrastructure projects' financial viability using different methods (Pantelias and Zhang, 2010). There are also some toolkits developed by the Public–Private Infrastructure Advisory Facility and the World Bank, which are available online to analyze a project's viability (Public–Private Infrastructure Advisory Facility [PPIAF], 2016). From the lenders' point of view, the project's investment risk and financial viability depend on the ability to repay the issued debt, which corresponds to the relationship between the project's costs and revenues during its life cycle. Cover ratios are the criteria that measure the project's ability to repay the debts as they fall due. *ADSCR* and loan-life cover are two of the most commonly used cover ratios. In practice, according to the perceived 'riskiness' of a project, the minimum acceptable cover ratios are determined by the lenders. They have to be fulfilled at all times for the project to be ultimately financed.

However, most of these methods are deterministic; they return a closed-form solution indicating the project's viability and cannot assess the potential investment risks. Risk assessment provides a way to estimate the probability that a project will meet its budget and time goals. Traditional risk assessment uses deterministic risk analysis based on single-point estimation and provides discrete outcomes but no information on the outcome's likelihood. Stochastic risk analysis integrates the range of possible values for each of the variables in the analysis. It provides the outcomes and their likelihood, according to various combinations of different input data with different values, reflecting the effects of risks on the outcome more intuitively. Different stochastic risk analysis techniques include the Bayesian algorithm, Monte Carlo experiments (MCE), and the fuzzy logic method. This section formulates a flexible framework to perform stochastic risk analysis with *Monte Carlo experiments*.

6.11.2 Simulation Experiments

Monte Carlo experiments is a widely applied computational algorithm based on a repeated random sampling process to calculate numerical results. This method uses parameters that can reflect the probability density function of variables as inputs, and the repetitive calculations take the randomly selected combinations of the inputs into consideration. The simulation outputs are the results, which are presented in a cumulative density function or probability density function. Specific weaknesses exist in this technique: it can be time-consuming if too many input variables are involved and excessive iterations are generated. The simulation may also produce a large variance.

To run *Monte Carlo experiments* to conduct stochastic risk analysis, practitioners need to identify the input variables and their associated distributions, which can usually be obtained from historical data or empirical assumptions. With the variables and associated distributions as the inputs, a sampling process is usually employed to simulate output variables' distributions with a prespecified number of iterations. Risks are assessed by comparing the output distributions with each output variable's targeted values; for example, the output distribution's proportion is less than the targeted value. *Monte Carlo experiments* has been used in various areas to conduct a stochastic risk analysis. Ridlehoover applied *Monte Carlo experiments* and risk analysis to help solve a facility location problem (Ridlehoover, 2004). Arnold and Yildiz conducted an economic risk analysis of decentralized renewable energy infrastructure using *Monte Carlo experiments* (Arnold and Yildiz, 2015). Da Silva Pereira et al. presented a method based on *Monte Carlo experiments* to estimate the behavior of economic parameters in power generation (da Silva Pereira et al., 2014). Carrasco and Chang used *Monte Carlo experiments* and risk assessment for ammonia concentrations in wastewater effluent disposal (Carrasco and Chang, 2005). Au et al. used advanced *MCS* in conducting compartment fire risk analysis (Au et al., 2007). *Monte Carlo experiments* has also been applied by rating agencies (e.g., Standard & Poor's and Moody's) to evaluate investment and credit risks at the country and industrial levels (Fender and Kiff, 2004).

Although *Monte Carlo experiments* has been widely used to evaluate infrastructure projects, a review of the literature found few studies on assessing specific 3Ps highway projects' investment risk. This chapter aims to apply this sophisticated technology to conduct a stochastic risk analysis to assess the tolled 3Ps highway projects' investment risks numerically. *NPV*, *IRR*, and *ADSCR* (for senior bank debt and combined debts) are selected as the outcomes. The outcome values, as well as their probability distribution, are calculated and analyzed. The results are compared with the deterministic values (no-risk scenario), which should provide additional information to facilitate public agencies, the private sector, and lenders in making more insightful decisions on the project's financial viability. This method can be applied to different scenarios in which different combinations of risks are considered.

6.11.3 *Pathway*

In general, typical project financing risks include construction risk, operational risk, supply risk, offtake risk, prepayment risk, political risk, currency risk, authorization risk, and dispute resolution risk. Studies have identified that the investment risk is a function of individual project risk, competitive risk, and market risk (Seneviratne and Ranasinghe, 1997; Javid and Seneviratne, 2000). Individual

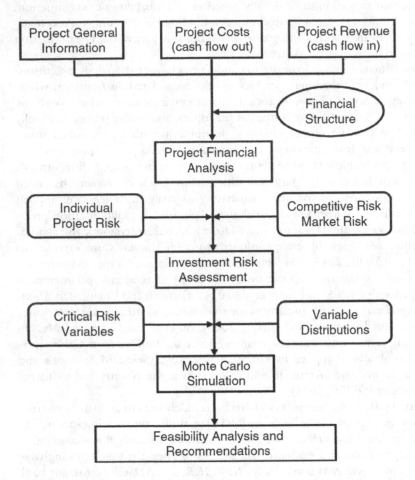

Figure 6.4 Methodological framework.

project risks can arise from inexperienced project contractors who may provide inaccurate and unreliable cost estimates and work-plan schedules. Competitive risks are caused by insufficient analysis before the project is undertaken, leading to imprecise infrastructure-generated revenues and inaccurate estimates of the project's market shares and its competitors. Market risks are the risks of losses in positions arising from movements in market prices. The methodological framework of this chapter is presented in Figure 6.4.

More specifically, to illustrate the method, several types of investment risks are considered in this chapter:

- Individual project risks, in which the risks in initial construction cost, initial operating cost, and maintenance and rehabilitation cost are considered.

- Competitive risks, that is, the risks in initial (AADT[2]) when the traffic volume that will use the toll highway facility and the traffic growth rate are predicted.
- Market risks, the senior bank interest rate, and the initial estimated inflation rate.

The variable values will fluctuate around the expected mean value rather than a deterministic number because of these risks. In this chapter, the effect of risks is simulated by setting an appropriate coefficient of variation (CV) and the reasonable probability distribution of the variables.

The equations to calculate NPV, IRR, and ADSCR are presented in Equations 6.4 through 6.7, respectively (Public–Private Infrastructure Advisory Facility, 2016):

$$NPV = \sum_{i=i_0}^{i_0+N} \frac{TCF_{i-i_0}}{(1+r)^{i-i_0}} \tag{6.4}$$

TCF_{i-i_0} = total cash flow for a year $i - i_0$
r = appropriate discount rate
i_0 = project base year
N = end year of concession

Furthermore, the total cash flow can be calculated as follows:

TCF = total revenues – total costs

= operating revenues + other revenues – construction cost

operation cost – maintenance cost – rehabilitation cost $\tag{6.5}$

repay debts – other costs

For the IRR,

$$\sum_{i=i_0}^{i_0+N} \frac{TCF_{i-i_0}}{(1+IRR)^{i-i_0}} = 0 \tag{6.6}$$

For the average debt service coverage ratio, the expression of the average debt service coverage ratio for a year $i - i_0$ is as follows:

$$ADSCR_{i-i_0} = 1 + \frac{TCF_{i-i_0}}{\sum_{j=1}^{J}(\text{principal repayment} + \text{interest payment})_j^{i-i_0}} \tag{6.7}$$

Where j is the jth debt service, and J equals total debt services.

6.12 A Report

A case study was carried out to illustrate the proposed method and demonstrate how it could serve as an analysis tool to help all stakeholders make decisions. The presented case study pertains to a real 3Ps highway toll road concession agreement, specifically a section (P12) from the Trans-Texas Corridor, a megaproject planned for Texas's operation. Although existing concessions were allowed to continue, the state legislature stopped developing the Trans-Texas Corridor and placed a two-year moratorium on new 3Ps projects due to specific issues (KWTX, 2005; KXII, 2006; Taylor, 2007, p. 17).

Since the investment risks have been identified, the coefficient of variation and distribution were assigned to the variables to simulate the risk. coefficient of variation is a dimensionless parameter used to describe non-deterministic variables' variability, which equals the standard deviation divided by its expected mean value. Based on some of the unsuccessful experiences, it is likely that agencies might provide inaccurate estimates on the cost and revenue information (Cuttaree, 2008; Minnesota Department of Transportation, online reference, 2016). In practice, one can identify the statistical coefficient of variation by translating massive historical cost- and revenue-related data through appropriate models or basing them on assumptions supported by empirical analysis. Skitmore and Ng analyzed 29 projects and showed that the coefficient of variation of total project cost ranged from 15.90% to 31.29%, with an average cost of 22.06% (Skitmore and Ng, 2002). Based on that finding, the initial construction cost coefficient of variation is set to be 20% in this case study. Wright et al. studied 21 sites in Florida and found that the coefficient of variation of Average Annual Daily Traffic ranged from 8% to 22% (Wright et al., 1997). Other studies have indicated similar ranges (Aunet, 2000; Turner et al., 1998). Therefore, the initial Average Annual Daily Traffic and annual traffic growth rate are set with a coefficient of variation of 15% and 10%, respectively. References on financial management also provided insightful information on various economic rates (Khan and Jain, 2005; Brigham and Ehrhardt, 2013). The coefficient of variation of the inflation rate and discount rate are set at 10%, and the interest rate of senior bank debt has a coefficient of variation of 5%.

Regarding the variable distribution, according to various previous studies, the cost risk variable and market risk variable are assumed to have a lognormal distribution, while the initial Average Annual Daily Traffic and annual traffic growth rate are placed as normally distributed (Baker and Trietsch, 2013; Wall, 1997; van Haastrecht and Pelsser, 2011; Piyatrapoomi et al., 2005; Miltersen et al., 1997; Marques and Berg, 2011). The coefficient of variation and distributions of each variable can be changed and customized depending on the available data and practitioners' preferences. The project P12 information is obtained from the original master development plan and summarized in Table 6.4 (Pantelias, 2009).

The distributions of initial construction cost and Average Annual Daily Traffic are illustrated as two examples of the input variables presented in Figure 6..5.

Table 6.4 Project P12 Information

Project parameter	Mean	CV (%)	Comments
General information			
Concession period	50 years	NA	
Construction period	5 years	NA	
Project length	57.0 mi	NA	
Number of lanes per direction	3	NA	Including shoulder
Project cost: initial construction cost (C_0)	$822,330,824	20	Initial construction estimate = design cost + ROW cost + structure cost Lognormal distribution
Operation cost	3.50%	NA	As a percentage of C_0, the annual cost
Rehabilitation cost	0.60%	NA	As a percentage of C_0, the annual cost
Annual price escalation rate	2.5%	10	Same as the inflation rate
Traffic and revenue			
AADT	24,278 vehicles	15	Initial estimate, normal distribution
Cars	65%	NA	
Trucks	35%	NA	
Annual traffic growth	6.5%	10	Normal distribution
Average trip length	30 mi	NA	
Average transaction per trip	1.3	NA	
Average transaction cost	$0.15/veh/transaction	NA	Per vehicle per transaction
The toll rate for cars	$0.152/car/mi	NA	Per car per mile
The toll rate for trucks	$0.585/truck/mi	NA	Per truck per mile
Annual toll rate growth	2.5%	10	Same as the inflation rate
Economic variables			
Inflation rate	2.5%	10	Initial estimate, lognormal distribution
Discount rate	12%	10	Target value for private sectors
Financing variables Construction capital draw			
Year 1	20%	NA	
Year 2	20%	NA	
Year 3	20%	NA	
Year 4	20%	NA	
Year 5	20%	NA	

(*Continued*)

Table 6.4 Continued

Project parameter	Mean	CV (%)	Comments
TIFIA loan	33%		As a percentage of total construction costs
Interest rate	5.10%	NA	
Grace period	11 years	NA	
Payback period	35 years	NA	
Payment terms			Interest plus principal in equal installments after the end of the grace period, minimum principal payment of $1,000,000
Senior bank debt	47%		As a percentage of total construction costs
Interest rate	5.55%	5	Lognormal distribution
Grace period	5 years	NA	
Payback period	40 years	NA	
Min. ADSCR	1.75 ×	NA	Required and targeted value
Payment terms			No payments during the grace period, interest plus principal after the end of the grace period
Combined debt min. ADSCR	1.10 ×	NA	Required and targeted value
Developer's equity	20%	NA	As a percentage of total construction costs

Note: NA = not applicable; ROW = right-of-way; veh = vehicle; TIFIA = Transportation Infrastructure Finance and Innovation Act.

Based on Equations 6.4 to 6.7, *Monte Carlo experiments* was conducted with the software @RISK developed by Palisade Corporation. Technically speaking, the greater the number of iterations performed, the better the precision of the method. The results will converge after a certain number of iterations. To better reflect the investment risks and obtain more accurate results, 10,000 iterations were performed. Risk simulation results for NPV, IRR, and average debt service coverage ratio are presented in Figure 6.6(a) to (d), respectively.

Consequently, the relative frequency (left y-axis) and the cumulative percentage curve (right y-axis) are provided for each outcome from the simulation results rather than a deterministic value. The larger the relative frequency, the more likely it is that the outcome will take place. The decision-making process is conducted by adjusting the delimiters in the analysis tool. Agencies can adjust the positions of delimiters to analyze the probability of achieving their objective

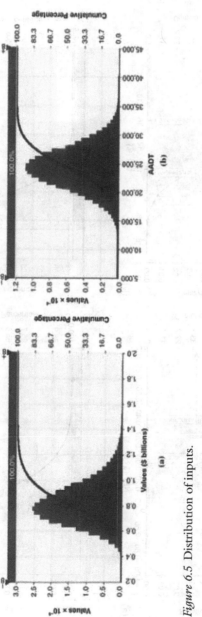

Figure 6.5 Distribution of inputs.

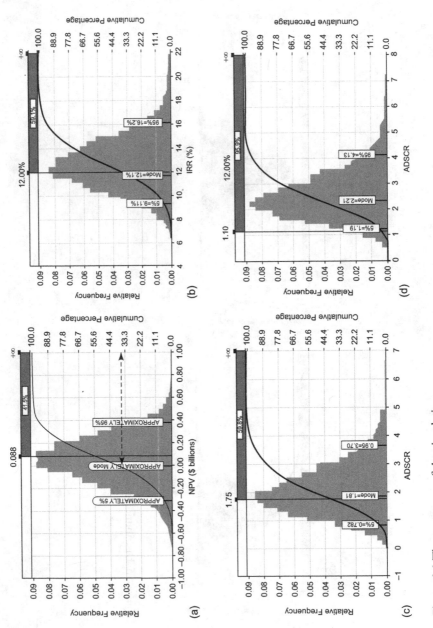

Figure 6.6 The outcomes of the simulation.

returns. Take Figure 6.6 as an example: if the target net percent value is $88 million, there is a possibility of 41.5% making the net percent value equal to or larger than the expected value.

On the other hand, there is a probability of 510.5% of failing this target. The output also indicates that the net percent value has a mean of $45.8 million, and there is a 36.1% probability of a negative net percent value. Figure 6.6(b) shows the behavior for internal rate of return. The most likely value, with an 8.4% probability of occurrence, is about 12.1%. There is a 59.1% probability of obtaining an IRR that is equal to or above 12%. Since 12% is the target value for the private sector (Table 6.4), there is a 40.9% probability of failing the target because of the risks. Based on Figure 6.6(c) and (d), debtors could have an overall estimation of the project's ability to meet the annual debt payment. According to Table 6.4, the minimum ADSCRs for senior bank debt and combined debts are 1.75 and 1.10, respectively. Figure 6.6(c) shows that the project's probability of meeting the senior bank debt's average debt service coverage ratio requirement is 59.8%. Figure 6.6(d) suggests a 95.9% probability that the project can meet the combined debts' average debt service coverage ratio requirement. Those findings indicate that the project can be considered to repay the combined debts reliably. Private investors and senior banks should be aware of and make prudent decisions concerning these investment risks. Furthermore, the results using deterministic values are calculated, and both results are summarized and listed in Table 6.5.

According to Table 6.5, the mean values from risk simulation are close to those from the deterministic analysis. Besides, all t-statistic values are less than 1.96, indicating that the deterministic and stochastic analyses are not significant at a 95% confidence level. The deterministic values cannot reflect the potential investment risk, whereas the stochastic analysis provides the outputs' distributions, making the investor more insightful. In the deterministic (no-risk) scenario, all four indexes indicate that the project is profitable and feasible. However, investors shall be aware that there is a probability of suffering a deficit when the investment risks are considered.

It also can be seen that the failure probabilities for NPV, internal rate of return, and ADSCR for senior bank debt are high. The reason is that this chapter involves too many risk factors and only toll-generated revenues for illustration

Table 6.5 Results Comparison

Evaluation method	NET PERCENT VALUE	IRR	ADSCR for senior bank debt	ADSCR for combined debts
Deterministic analysis	$45.4 million	12.4%	1.94	2.38
Risk simulation				
Mean	$45.8 million	12.5%	2.03	2.48
Standard deviation	$272.47 million	2.46%	1.015	0.992
Probability of failing (%)	36.1	40.9	40.2	4.1
t-statistic	0.002	0.041	0.089	0.101

purposes. Generally, the more uncertainties there exist in a project, the higher the failure probability will be. The failure probabilities will be reduced if only one or some of the risks are involved. Considering the initial construction of $822 million, the target internal rate of return of 12%, and the various risks involved in this case study, the project is acceptable. Still, it may not be so attractive to investors with so many risk factors. For the lenders, the project can reliably repay the combined debts. Although it might be unbalanced as some factors are not considered in the simulation process, the case study illustrates how the risk simulation conception works and the investment risk assessment process. The agencies can assess certain project-based risks with this method and make the decision practically. Minimization approaches for each type of risk are suggested for stakeholders, as listed in Table 6.6 (Marques and Berg, 2011). An input variable with lower risk should concentrate around its mean value with a smaller CV value.

3Ps have been widely applied over the years, and individual lessons have been learned. Both the public and private sectors should pay attention to common problems during the 3Ps process, such as risk and reward allocation, public- and private-sector capacity evaluation, project governance, and affordability issues. This section presents a methodological framework for assessing the investment risks associated with 3Ps toll highway projects. Instead of a deterministic value, the variables' distributions are used as inputs to simulate the risks. NPV, internal rate of return, and ADSCR are commonly used criteria for investors and lenders to evaluate the project's financial viability and ability to repay the debts. A detailed case study of a section (P12) from the Trans-Texas Corridor was conducted. The results are interpreted so that quantitative information was provided to agencies to establish investment decision criteria. The stochastic risk analysis helps agencies to make more insightful decisions.

The proposed framework works in conjunction with other models, such as cost and revenue prediction models. The procedures are flexible for practitioners to adopt. The coefficient of variation and distribution of an input variable are not limited to a specific value or form. Instead, practitioners can customize input variables associated with their coefficient of variation and distribution according to the available data or professional judgment in practice. Multiple simulations can be run with different coefficient of variation and distributions to assess different levels of risks, allowing 'what-if' scenarios to be studied. The method can serve as a basis for future research. Future research can embark on a few directions and improve the presented framework and corresponding results. They include simulating the cost risk at the line-item level, namely, modeling operating cost, maintenance cost, and rehabilitation cost as independent costs with individual coefficient of variation and distribution, and incorporating uncertainties in willingness to pay and toll elasticity in the revenue risk evaluation.

6.13 Risk of Corrupt Activities in 3Ps

Multimillion-dollar water infrastructure projects carry some of the most considerable corruption risks in the sector linked to the procurement of civil works and

Table 6.6 Minimisation Approaches for Each Type of Risk

Risk	Minimization approaches
Capacity	Increase studies accuracy; cost-benefit analysis
Competition	Sensitivity analysis; public disclosure of indicators
Conception	Careful selection of project designers; realism in studies planning; auditing studies and projects; contracts with premiums and fines
Construction	Strict management; fixed-price contracting; insurance contracting
Collection	Sensitivity analysis; service interruption; making payment easier; customers and collection management
Demand (consumption)	Sensitivity analysis; sensitizing actions; making payment easier
Environmental	Sensitizing actions; supervision and research; pressure on the authorities
Expropriation	Experienced work teams; project compatibility; fixed-price contracting
Financial	Long-term financing; hedging policies; backup funding (bank account)
Force majeure	Mostly protected; insurance contracting
Inflation	Indexation of revenues to inflation; fixed-price contracting; forward contracts
Legal	Protected by contract
Maintenance or repairs	Association to specialized companies; fixed-price contracting; insurance contracting
Operation	Association to specialized companies; fixed-price contracting; insurance contracting
Performance	Systematic control; fixed-price contracting
Planning	Careful selection of project designers; increase detail in studies
Public contestation	Sensitivity analysis; public disclosure of indicators
Regulation	Keep up with the international trend; systematic control of performance; benchmarking policies
Technological	Contracts with warranties; insurance contracting
Unilateral changes	Protected by contract

related design, supply, and consultancy services. The potential for grand corruption in big dam projects and upgrading urban water and sanitation systems can be so significant as to skew policymaking towards the most lucrative investments. 'White elephants,' such as overly sophisticated new wastewater treatment plants, may come at the expense of maintaining existing assets and more appropriate lower-cost technologies and approaches.

The water sector's big-ticket items are urban water and sewerage investments (including new or upgraded supply, sewerage, drainage networks, storage reservoirs, water, and wastewater treatment plants) and surface water storage and inter-basin transfers (dams, civil works, and related resettlement). Donor support for constructing large dams fell towards the end of the last century – at least partly due to concerns about social and environmental impacts (World Commission on Dams, 2000). But the World Bank has more recently put significant dam investments back on the agenda, pledging to 're-engage with high-reward high-risk hydraulic infrastructure' (World Bank, 2004). East Asia is increasingly investing in many new major infrastructure projects in Africa, including certain dams. Over half of OECD-DAC countries' support to the water sector in 2006-07 was for large water supply and sanitation systems.[3] In an era where developing water governance is a key priority on the development policy agenda, infrastructure investments strongly dominate donor funding in water and sanitation compared to 'soft' support for policies, legal systems, and capacity building (World Water Assessment Programme, 2009).

Large infrastructure projects are high on the agenda for several reasons. Urbanization means most urban systems need to be significantly expanded, and climate change and water security issues have encouraged new interest in water transfer schemes. Ever-growing energy demands also motivate new hydropower investments. In dams, there is the recognition that Africa still has to address deficient levels of water storage. In industrialized countries, water storage ensures reliable water sources for irrigation, water supply, and hydropower and provides a buffer for flood management. However, countries in Africa store only about 4% of annual renewable flows, compared with 70–90% in many developed countries (World Water Assessment Programme, 2009).

Due to their size alone, major infrastructure projects potentially offer immense rents for corrupt politicians and officials. The construction section is ranked globally as one of the sectors most vulnerable to corruption (Transparency International, 2005). Competitive tender processes – the best available norm for procuring products and services for major projects – have their strengths and encourage a winner-take-all mentality. Some firms risk paying bribes to gain a benefit over competitors. Major projects' complex and international nature includes potentially international sources of finance, consultants, and contractors. This means that corruption can also cross borders, involving countries with different laws and business cultures.

In Eastern Europe and Central Asia, construction firms have reported paying an average of 7% of government contract values in bribes to win bids or alter contract terms (Kenny, 2006). Such corruption raises the price of infrastructure and can also reduce the quality of and economic returns to infrastructure investment (Kenny, 2007). The challenge for donors is to ensure that development aid strengthens governance in the sector rather than further weakening it by providing an approach for corrupt elites to extract rents and enrich themselves. Major infrastructure investments provide a litmus test for the quality of donor support to the water sector.

6.14 Corruption and urban infrastructure

Grand corruption. Involving a relatively small number of individuals but containing large amounts of money and an abuse of discretionary power – this type of corruption is the most significant risk in major infrastructure projects. Corruption may extend to such a high level (state capture) that national policy is influenced and, in turn, motivates decision-making that favors activities that provide the most potential for corruption. For example, large and expensive infrastructure investments may be deliberately preferred in policy over smaller, decentralized systems and lower-cost technologies.

Corrupt acts always require two sectors. At this scale, corruption is likely to be among public officials and their colleagues in other departments (public–public interactions) or between public officials and private companies providing materials, equipment, and services. Grand corruption typically involves politicians, senior officials, and higher-level technical staff (González de Asís et al., 2009). Because of its size and sensitivity, the parties involved will significantly conceal their actions. A common practice of private companies, for instance, is to employ representative agents with a brief to secure contracts and provide a veneer of respectability and distance when bribery is involved.

6.15 An Evidence

One of the most widely known grand corruption cases in the section – celebrated for its David and Goliath nature of a small government ultimately holding major international companies responsible – was uncovered in the Lesotho Highlands Water Project construction.

> Massive corruption was explored in the LHWP in 1999 when more than 12 multinational firms and consortiums were found to have bribed the Chief Executive (CE) of the project. After the CE himself was found guilty, three major construction firms were put in the dock; thus far, three have been found guilty and charged, and one[4] has been debarred at the World Bank.
>
> (Stålgren, 2006)

This was a prestigious project and the largest international water transfer at the time, storing and distributing water from the Lesotho Mountains to the Gauteng industrial heartland of South Africa. High volumes of water were involved: 750 million cubic meters per year. And high volumes of money: the expected income on payments for water via South Africa was sufficient to cover Lesotho's foreign debt at the time.

Unfortunately, some major construction contracts were awarded fraudulently. The chief executive was prosecuted after he appealed against his dismissal over an initial investigation into the suspected misuse of cars and expenses (Earle and Turton, 2005). It was then discovered that foreign companies had paid bribes (totaling over USD 1 million over nine years) through their agents into international accounts. The companies involved strongly resisted prosecution by name changes and takeovers to escape liability (Darroch, 2007). However, Lesotho mounted a successful prosecution despite its high cost and the limited support it received from financiers and the international community. Finally, the chief executive received 15 years in jail for bribery, and major international companies and their agents were also convicted. The successful prosecution of both givers and takers, in this case, set some crucial precedents: (1) that jurisdiction can be taken where the effect is felt (e.g., in Lesotho), (2) that the giver and taker of bribes are equally accountable, (3) bribes are still illegal even if not acted on after bribe agreement is reached, and (4) courts could gain access to Swiss bank accounts.

The Lesotho trial showed that multinational companies could be held to account even by a small country with limited resources. However, the judicial system's limited capacity in many countries, the high cost of enforcing accountability through the courts, and the lack of international support for such prosecutions are significant concerns. Considering that the trial risks being one of a kind, it is asked that the case be "viewed through an international lens, with the international community taking responsibility for the role it can best play in the future" (Darroch, 2007).

Potential forms of corruption in major infrastructure projects include (González de Asís et al., 2009):

- **Bribery**: the giving of some form of profit to unduly influence an action or decision.
- **Collusion**: an arrangement among two or more parties designed to achieve an improper purpose, such as when bidders for contracts agree between themselves on prices and 'who should win.' This may involve paying bribes to public officials to 'turn a blind eye.'
- **Embezzlement and theft**: the taking or conversion of money, property, or other valuables including, for instance, the diversion of public funds to a personal bank account.
- **Fraud**: the use of misleading information to induce someone to turn over money or property voluntarily, for instance, via misrepresenting the number of people in need of a particular service.
- **Extortion**: involving coercive incentives such as the use of threats of violence or the exposure of damaging data to induce cooperation. Officeholders may be either the instigators or the victims of extortion.

One hotspot for these risks is procurement. In the water sector (mainly run as a public service), there are high requirements for the products and services from the private party. The landscape of these public–private interactions has been changing, with privatization widely promoted over the past decade or two. This has provided new openings for corruption through contracting out, concessions, and privatization in the context of inadequate regulation (Hall, 1999). Davis (2004), in her survey of corruption in the water sector in South Asia, reports that, for this type of corruption, 'the value of kickbacks paid was fairly consistent … between 6% and 11% of the contract value, on average.' Another related corruption hotspot is the unsatisfactory completion of projects, frequently including collaboration between supervising consultants and contractors.

In the construction of large dams, corruption in the resettlement of displaced people has also been a major problem, along with other forms of corruption in planning and construction. Marginalized re-settlers may fail to receive compensation and development benefits to which they are entitled. The largest dam-building project – the Three Gorges Dam in East Asia – gave rise to the most extensive corruption scandal, with officials embezzling an estimated USD 50 million from funds set aside to resettle over a million people. In this case, compensation was handled by a decentralized agency and provided an opportunity for local governments to misuse funds. Accounting and auditing systems were subsequently strengthened with more oversight provided, but insufficient transparency and participation are still challenging, and problems persist.

6.16 Combatting Against Corruption in 3Ps

During the 1970s and 1980s, environmental effect assessments became a mainstream tool to consider and mitigate projects' potential environmental impacts. A crucial recommendation is that corruption risk assessment should also become standard practice in significant water-sector projects. Analyzing potential corruption risks and putting preventative measures in place is easier and much more cost-effective than trying to clean up corruption after it becomes established. Although universally adopted guidelines are not yet available, the necessary tools for such assessments are now available. Sector risk assessments and water integrity scans highlighting potential corruption hotspots and early warning signs and preventative measures are currently being trialed in countries like Ethiopia with World Bank support and Uganda with the Water Integrity Network. Such assessments must lead to strategies and action plans that build upon the risks identified. For example, Mozambique has recently embarked on developing a water-sector anti-corruption strategy (focusing on the National Department of Water).

International competitive bidding may reduce collusion opportunities and greatly extend procurement processes and significantly add costs (Kenny, 2007). However, an essential step in minimizing risks and the best available tool, standard tendering, and procurement procedures will not always prevent corruption and might even make things worse in some situations. Elements such as the lengthy nature, high cost, intense competition, and complicated administration of bid processes (coupled with the bonus to be paid to the bid manager) encourage a winner-takes-all mentality where unsuccessful bidders stand to lose a lot. With few ways to improve a real competitive advantage[5] (since technical designs are commonly proscribed), profit-driven firms may resort to bribes to win contracts (Campen, 2009). Strengthening procurement systems requires ongoing support in capacity building. In high-risk countries and projects, many donors insist on outsourcing procurement to international companies. This may be necessary but is not a long-term solution to help build capacities and oversight mechanisms within the government.

One strategy which demonstrates that preventative measures are available to prevent corruption in major projects is integrity pacts (González de Asís et al., 2009). Developed via Transparency International in the 1990s to safeguard public procurement from corruption, integrity pacts aim to decrease the chances of corrupt practices during procurement through a binding agreement between the agency and bidders for particular contracts. They are intended to decrease the high costs of corruption in public procurement. The pact is made between a procurement agency (generally governmental) and bidders for separate agreements. It enables companies to abstain from bribing by assuring them that their competitors will also refrain from paying bribes. Public agencies also pledge to prevent corruption, including not seeking to bribe. Integrity pacts have already been implemented in several countries (including Argentina, Colombia, and Mexico) in infrastructure projects in the water and sanitation sectors. However, scaling up such approaches remains a challenge and requires donors, governments, the private sector, and facilitators like NGOs and professional associations. In countries where the enabling environment is not yet conducive for such pacts at the national level, piloting such strategies sub-nationally (where there is a commitment from a city or district government) could be a way forward and may set a favorable example.

6.17 A Case Study from Mexico

In Mexico, the 'social witness' is a representative of civil society who acts as an external observer in the procurement process (OECD, 2006). To promote transparency, reduce the risk of corruption, and develop procurement's overall efficiency, this innovative practice in integrity pacts has been utilized for several years, following Transparencia Mexicana's recommendation. The social witness – a highly honorable, recognized, and trusted public figure independent of the parties encountered in the process – makes recommendations during and after the procurement procedure and provides public testimony. Regulations demonstrate criteria for participation of the social witnesses in procurement, and a list of registered social witnesses is published on the Ministry of Public Administration website (see www.funcionpublica.gob.mx/unaopspf/unaop1.htm). Transparencia Mexicana, for instance, acted as the social witness for the procurement of sewerage treatment services by the Municipality of Saltillo in 2004, an agreement worth almost USD 5 million. They followed each stage of the procurement process, attended meetings, and provided advice to the municipality. They introduced a signed summary statement on completion of the procurement testifying that the process was proper and describing what occurred at different stages (for instance, why certain bidders failed). It was also explained why the agreement was awarded to the successful bidder.

6.18 Donors Prospects

Donors can strengthen their accountability systems by developing access to documents, taking action against corrupt staff, and blacklisting corrupt project partners (O'Leary and Stalgren, 2008). They also need to provide positive encouragement and support to their staff to combat corruption. Finding out about corruption and acting on these data can cause issues. Civil servants risk not being promoted if they uncover corruption in their programs, and as a result, many would prefer to keep quiet.

Donors can also strengthen the anti-corruption components of water programs, promote civil society capacity and media improvement (important in encouraging recipient governments to be accountable to citizens and donors), and put pressure on governments to implement anti-corruption plans and strategies within the sector. Positive roles that donor countries can play in major projects include encouraging corruption prevention at the design stage, addressing corruption risks in the implementation itself (especially contracting and renegotiations), and post-corruption follow-up, including support to prosecutions (a key lesson from the Lesotho Highlands experience).

6.19 Kickbacks

Anyone involved in bidding for and executing government engineering projects will be aware of their potential for corruption. According to a Transparency International survey, public works and construction were ranked as the most corrupt industry worldwide (Transparency International, 2002).

Why is this so? What is it about infrastructure projects that render them particularly susceptible to corruption, and what might be done to limit such corruption? Is there a role for the civil engineering profession in facilitating efforts to control corruption?

Corrupt kickbacks are easy to hide in construction contracts, and the competitive nature of many bidding processes encourages firms to try to circumvent them through payoffs. Also, once the contract is written, officials may seek to extract payoffs from the contractor. Unscrupulous contractors have an incentive to pay bribes that permit them to cut corners to increase profits. In a transparent, competitive bidding process for a standardized product, corruption would not be possible so long as the head of government sought a clean process, either out of moral scruples or because he or she fears a loss of public support if corruption is revealed. In a simple process, the most efficient firm would be the low bidder; if another firm tried to bribe officials, that firm would have to submit a bid high enough to cover the bribe's cost. The discrepancy between the high-winning bid and the lower bid would be evident to all and would have negative political consequences. Therefore, the only firm that could get away with paying a bribe would be the most efficient. However, that firm would have no incentive to pay a bribe because it can get the contract without that expense (Rose-Ackerman, 1978).

Notice the conditions that produce this result

a. A standardized set of specifications
b. Competition between firms
c. Transparency of bids
d. A political leader who will suffer politically from revelations of corruption

Consider each in turn. If any of them is violated, then corruption can occur, not just during the bidding but at other points in the procedure.

First, many civil engineering projects are not fully specified *ex-ante* by the contracting state. Hence, bidders both offer a price and specify aspects of the projects they think will persuade the government to select them even if their price is higher. So long as the costs of idiosyncratic provisions are not common knowledge, kickbacks can be hidden in the special provisions that bidders propose. Bribes can be hidden in those parts of the contract, and the government's choice of such a bidder does not immediately signal that corruption has occurred.

Second, if there is little or no serious competition for a contract and no benchmark prices for the service in world markets, kickbacks can quickly be paid.

One common corrupt technique is to help the government draft the specifications so that your firm is the only qualified bidder. For example, consider a perhaps apocryphal story of a tropical African country that sought to purchase telephones that would function well if the temperature dropped below freezing. Only one firm could meet that specification.

6.20 Cartel

When a contractor has monopoly power, it may refuse to pay bribes because the government has no choice but to deal with that firm. Thus, if the firm's management and owners are otherwise independent of the state, they can resist. This is what happened in the United States. A former governor of Maryland orchestrated a corrupt contracting system. Still, one or two specialized bridge engineering firms were exempt because only they could fulfill specific necessary tasks (*New York Times*, 1973).

However, in very corrupt systems, a firm's monopoly power may facilitate corrupt deals between top officials and the firm at the public's expense. This was the case in states like Zaire and municipal governments operating in tight relationships with organized crime, as in parts of Italy (Rose-Ackerman, 1999).

If corruption does not have negative political consequences for officials, they can orchestrate corrupt systems that benefit them personally.

6.21 Offers Manipulation

In addition to individual kickbacks by bidders, corruption can facilitate bid-rigging. Thus, even if there appears to be vigorous competition for government contracts, this may be a sham that hides a system of sharing the work and keeping prices high. Although bid-rigging can occur without payoffs, officials are likely to be aware of firms' activities and must be bought off to keep them quiet. Even if officials are not corrupt and suspect bid-rigging, it may be challenging to combat, as evidenced by the US Army's efforts in South Korea. Suppliers who attempted to operate outside the bid-rigging system were threatened and intimidated by the cartel (Klitgaard, 1988; Della Porta and Vannucci, 1999). Nevertheless, especially in a municipal government with access to national or international firms, efforts to bring in competitors can be successful. For example, New York City bought in a national waste management firm to collect rubbish as a way to break organized crime control (Rose-Ackerman, 1999).

6.22 Conclusions and recommendations

How can such corruption be countered? Of course, the primary focus must be on the governments involved. They need to improve their bidding processes along the lines suggested by Transparency International (2005). Beyond that, however, they need to reexamine what they are purchasing. In the face of corruption, they should ask if they can shift to more standardized goods and services where national or international benchmarks exist and where firms' bids can be compared. For this to occur, they might consider doing more of the engineering and design work in-house so that contracts are made at a point where they can be standardized (Rose-Ackerman, 1999).

Turning to engineering firms, they can, of course, refuse to bid on projects in countries or sub-national governments where kickbacks are aggressively demanded. However, that strategy leaves the country's citizens at the mercy of firms willing to play the corrupt game. Engineering firms with a policy against paying bribes should consider bidding honestly for projects in such jurisdictions and then publicizing their problems and their bids' value. Furthermore, they might set conditions on their participation, demanding that all bidders implement internal anti-corruption controls, following the protocol developed by Transparency International (2005). Bidders would report suspected corruption cases to the prosecutors in the home countries of firms resident in countries that are parties to the Organization for Economic Co-operation and Development anti-bribery convention (Organization for Economic Co-operation and Development, 1997).

As for the professional engineering societies, they could compile a database of the costs of various standard projects under a set of common conditions. Because the technical and engineering data are often not proprietary for many joint civil engineering projects, this could be done without revealing trade secrets and could help international watchdog groups and domestic civil society organizations hold governments and their contractors to account. These actions would benefit firms that seek to do business honestly by providing circumstantial evidence of corruption in projects that vastly exceed benchmarks or are overly specialized and complicated for a particular population's needs. Such data would seldom provide evidence sufficient for legal enforcement actions. Still, its use could provoke debate and put the burden of proof on contracting authorities to justify their expensive choices.

- Major water infrastructure projects offer the most significant opportunities for corruption and should always be subject to corruption risk assessment. Such assessments ought to be mainstreamed via standard guidelines being adopted for environmental impact assessments.
- Where dams and other projects have significant social effects like the displacement of communities, transparent procedures for early engagement in planning and dealing with community complaints (including the role of an ombudsman) are essential, as are suitable communication strategies. Communities should be made partners in such developments.
- Donors and other actors with influence need to look out for the impacts of corruption, or potential corruption, on policy and decision-making. Specifically, close consideration should be given to why large-scale investments are prioritized over other alternatives, suggesting better value for money and more sustained access to water and sanitation.
- Procurement processes could be strengthened and made more effective through more incredible study and developed documentation and capacity building in this area. Piloting and scaling up promising approaches such as integrity pacts and social witnesses should also be a priority.

Notes

1 Suzhou is one stop along the high-speed railway between Shanghai and Nanjing.
2 Average Annual Daily Traffic
3 Member countries of the OECD Development Assistance Committee.
4 At the time of writing two firms have now been disbarred.
5 One way for companies to differentiate themselves from their competitors is to adopt and enforce an anti-bribery policy.

References

Abdel Aziz, A. M. (2007). Successful delivery of public-private partnerships for infrastructure development. *J. Constr. Eng. Manag.*, *133*(12), 918–931.

Acemoglu, D., & Robinson, J. (2012). *Why nations fail: the origins of power.* New York, NY: Crown Publishers.

ACG (2007). Performance of PPPs and traditional procurement in Australia. Final Report to infrastructure partnerships Australia, Sydney, Australia.

Ahadzi, M., & Bowles, G. (2004). Public–private partnerships and contract negotiations: an empirical study. *Constr. Manag. Econ.*, *22*(9), 967–978.

Ai, S., Du, R., Brugha, C. M., & Wang, H. (2016). Pointing to priorities for multiple criteria decision making—The case of a MIS-based project in China. *Int. J. Inform. Technol. Decision Making*, *15*(03), 683–702.

Aidt, T. S. (2003). Economic analysis of corruption: a survey. *Econ. J. 113*(491), 632–652.

Ajzen, I. & Fishbein, M. (1975). *Belief, attitude, intention, and behaviour: an introduction to theory and research.* Reading, MA: Addison-Wesley.

Ajzen, I. & Fishbein, M. (1980). *Understanding attitudes and predicting social behaviour.* Englewood Cliffs, NJ: Prentice-Hall.

Ajzen, I. (1985). From intentions to actions: a theory of planned behaviour. In: J. Kuhland & J. Beckman, eds. *Organizational behaviour and human decision processes* (pp. 11–39). Heidelberg: Springer.

Ajzen, I. (1991). The theory of planned behavior. *Organ. Behav. Hum. Decis. Process.*, *50*(2), 179–211.

Akbar, H. Y., & Vujić, V. (2014). Explaining corruption: the role of national culture and its implications for international management. *Cross Cult. Manag.: Int. J.*, *21*(2), 191–218.

Akerlof, G. & Kranton, R. (2000). Economics and identity. *Q. J. Econ.*, *115*(3), 715–753. URL: http://ideas.repec.org/a/tpr/qjecon/v115y2000i3p715-753.html

Akintoye, A., Hardcastle, C., Beck, M., Chinyio, E., & Asenova, D. (2003). Achieving best value in private finance initiative project procurement. *Constr. Manag. Econ.*, *21*(5), 461–470.

Alesina, A. et al. (2003). Fractionalization. *J. Econ. Growth, 8*, 155–194. URL: http://ideas.repec.org/a/kap/jecgro/v8y2003i2p155-94.html

Alfen, H., Kalidindi, S., Ogunlana, S., Wang, S. Q., Abednego, M., Frank-Jungbecker, A., Jan, Y. C., Ke, Y., Liu, Y. W., Singh, B., & Zhao, G. F. (2009). Public-private partnership in infrastructure development: case studies from Asia and Europe. Public-private partnership in infrastructure development: case studies from Asia and Europe, Bauhaus-Universität Weimar, Faculty of Civil Engineering, Chair of Construction Economics, vol 7, pp. 11–13.

Alkan, R. M., Erol, S., Ozulu, I. M., & Ilci, V. (2020). Accuracy comparison of post-processed PPP and real-time absolute positioning techniques. *Geomatics, Natural Hazards, and Risk, 11*(1), 178–190.

Almerighi, M. (1993, March 1), Il giudice dello scandalo petroli: non ripetiamo certi errori. *Il Corriere della sera*, 1° marzo 1993 - Pagina 4. Milano, Italy.

Ambraseys, N., & Bilham, R. (2011). Corruption kills. *Nature, 469*(7329), 153–155.

Ameyaw, E. E., & Chan, A. P. (2016). A fuzzy approach for the allocation of risks in public–private partnership water-infrastructure projects in developing countries. *J. Infrastruct. Syst., 22*(3), 04016016.

Amos, P. (2004). Public and private sector roles in the supply of transport infrastructure and services. Operational Guidance for the World Bank Staff, Transport Paper – 1. Washington, D.C.: The World Bank.

Amundsen, I. (1999). Political corruption: an introduction to the issues (No. CMI Working paper WPI 1999:7). *Human Rights*. Bergen, Norway: Chr. Michelsen Institute.

ANAC, A. N. A. (2007). Deliberazione - Risoluzione Del Consiglio Del 19 Dicembre 2007. RISOLUZIONE DEL CONSIGLIO del 19 dicembre 2002 sulla promo zione di una maggiore cooperazione europea in materia di istruzione e formazione professionale (2003/C 13/02).

Anand, V., Ashforth, B. E., & Joshi, M. (2004). Business as usual: the acceptance and perpetuation of corruption in organizations. *Academy of Management Executive, 18*, 39–53. http://doi.org/10.5465/AME.2004.13837437

Anastasopoulos, P., Florax, R., Labi, S., & Karlaftis, M. (2010). Contracting in highway maintenance and rehabilitation: are spatial effects important? *Transportation Research Part A: Policy and Practice, 44*(3), 136–146.

Angilella, S., Corrente, S., & Greco, S. (2015). Stochastic multiobjective acceptability analysis for the Choquet integral preference model and the scale construction problem. *European Journal of Operational Research, 240*(1), 172–182.

Angilella, S., Corrente, S., Greco, S., & Słowiński, R. (2016). Robust Ordinal Regression and Stochastic Multiobjective Acceptability Analysis in multiple criteria hierarchy process for the Choquet integral preference model. *Omega, 63*, 154–169.

Anti-Corruption Resource Centre. (2015). Glossary. available at: http://www.u4.no/glossary/.

Ariely, D. (2012). *The honest truth about dishonesty: how we lie to everyone-especially ourselves.* London, UK: Harper Collins.

Arino, A., de la Torre, J., & Smith Ring, P. (2001). Relational quality: managing trust in corporate alliances. *California Management Review, 44*(1), 109–131.

Arnold, U., & Yildiz, Ö. (2015). Economic risk analysis of decentralized renewable energy infrastructures–A Monte Carlo simulation approach. *Renew. Energy, 77*, 227–239.

ASCE (2009). *Report Card for America's Infrastructure*, 2009. http://www.asce.org/reportcard/2009/. Accessed July 28, 2016.

Asenova, D., & Beck, M. (2003). A financial perspective on risk management in public-private partnership, PUBLIC-PRIVATE PARTNERSHIPS, edited by Akintoye et al.

Ashforth, B. E., & Anand, V. (2003). The normalization of corruption in the organization. *Research in Organizational Behaviour*, 25, 1–52. http://doi.org/10.1016/S0191-3085(03)25001-2

Au, S. K., Wang, Z. H., & Lo, S. M. (2007). Compartment fire risk analysis by advanced Monte Carlo simulation. *Engineering Structures*, 29(9), 2381–2390.

Aunet, B. (2000). Wisconsin's approach to variation in traffic data, Wisconsin Department of Transportation, USA.

Authers, J. (2015). Infrastructure: bridging the gap. *Financial Times*, USA.

Auti, A., & Skitmore, M. (2008). Construction Project Management in India. *Int. J. Confl. Manag.*, 8(2), 65–77.

Axelrod, R. (1984). The *evolution of coo*peration. New York: Basic Books Harper Collins.

Ayal, S., & Gino, F. (2011). Honest rationales for dishonest behavior. In: M. Mikulinicer & P. R. Shaver (Eds.), *The social psychology of morality: exploring the cause of good and evil* (pp. 149–166). Washington, DC: American Psychological Association. doi:10.1037/13091-008

Azfar, O., Lee, Y. & Swamy, A. (2001). The causes and consequences of corruption. *ANNALS Am. Acad. Political Social Sci.*, 573(1), pp. 42–56. URL: http://www.jstor.org/stable/10.2307/1049014

Babalos, V., Caporale, G. M., & Philippas, N. (2012). Efficiency evaluation of Greek equity funds. *Res. Int. Bus. Financ.*, 26(2), 317–333.

Baizakov, S. (2008). *Guidebook on promoting good governance in public-private partnership*. United Nations Economic Commission for Europe.

Baker, K. R., & Trietsch, D. (2013). *Principles of sequencing and scheduling*. USA: John Wiley & Sons.

Ballesteros-Pérez, P., Skitmore, M., Pellicer, E., & González-Cruz, M. C. (2015). Scoring rules and abnormally low bids criteria in construction tenders: a taxonomic review. *Constr. Manag. Econ.*, 33(4), 259–278.

Bandura, A. (1999). Moral disengagement in the perpetration of inhumanities. *Personality Social Psychol. Rev.*, 3, 193–209. doi:10.1207/s15327957pspr0303_3

Bandura, A., Adams, N. E., Hardy, A. B., & Howells, G. N. (1980). Tests of the generality of self-efficacy theory. *Cognitive Therapy and Research*, 4, 39–66.

Baporikar, N. (2020). Dynamics in implementation of public private partnerships. *Int. J. Political Activism Engage. (IJPAE)*, 7(1), 23–53.

Bardhan, P. (1997). Corruption and development: a review of issues. *J. Econ. Literature*, 35, 1320–1346. Stable URL: http://www.jstor.org/stable/2729979

Bargh, J., & Chartrand, T. (1999). The unbearable automaticity of being. *Am. Psychol.*, 54(7), 462–479. http://dx.doi.org/10.1037/0003-066X.54.7.462

Barron, F. H., & Barrett, B. E. (1996). Decision quality using ranked attribute weights. *Manage. Sci.*, 42(11), 1515–1523.

Bassioni, H. A., Price, A. D. F., & Hassan, T. M. (2004). Performance measurement in construction. *J. Manag. Eng.*, 20(2), 42–50.

Batson, C. D. (2016). Moral motivation: a closer look. In: J. W. van Prooijen & P.AM. Van Lange (Eds.), *Cheating, corruption, and concealment: the roots of dishonesty* (pp. 463–484). Cambridge: Cambridge University Press.

Batson, C. D., & Powell, A. A. (2003). Altruism and prosocial behavior. In: T. Millon & M. J. Lerner (Eds.), *Handbook of psychology, volume 5: personality and social psychology* (pp. 463–484). Hoboken, NJ: Wiley.

Beatham, S., Anumba, C., Thorpe, T., & Hedges, I. (2004). KPIs: a critical appraisal of their use in construction. *Benchmarking: An Int. J.*, *11*(1), 93–117.

Beck, P. J., & Maher, M. W. (1986). A comparison of bribery and bidding in thin markets. *Econ. Lett.*, *20*(1), 1–5.

Belassi, W., & Tukel, O. I. (1996). A new framework for determining critical success/failure factors in projects. *Int. J. Proj. Manag.*, *14*(3), 141–151.

Beria, P., & Grimaldi, R. (2011). An early evaluation of Italian high-speed projects. *Tema*, *4*(3), 15–28.

Bertelli, A. M., Mele, V., & Whitford, A. B. (2019). When new public management fails: infrastructure public-private partnerships and political constraints in developing and transitional economies. Forthcoming, Governance.

Bicchieri, C., & Rovelli, C. (1995). Evolution and revolution: the dynamics of corruption. *Rationality Society*, *7*(2), 201–224. http://doi.org/10.1177/10 43463195007002007

Bicchieri, C., & Xiao, E. (2009). Do the right thing: but only if others do so. *J. Behav. Decision Making*, *22*(2), 191–208. http://doi.org/10.1002/bdm

Binmore, K. (1992). *Fun and games: a text on game theory*. D.C. Heath.

Biondani, P. (2000, May 17), Alta Velocità, condannati Necci e Pacini Battaglia. *Corriere della sera*, Italy, Milano.

Black, F., & Scholes, M. (1973). The pricing of options and corporate liabilities. *J. Political Econ.*, *81*(3), 637–654.

Blank, F. F., Baidya, T. K., & Dias, M. A. (2009). Real options in public private partnership: case of a toll road concession. In: 13th annual international conference on real options. Portugal: Real Options Group.

Blau, A. (2009). Hobbes on corruption. *History Political Thought*, *30*, 596–616.

Bologna, R., & Del Nord, R. (2000). Effects of the law reforming public works contracts on the Italian building process. *Build. Res. Inf.*, *28*(2), 109–118.

Bosworth, R. J. B. (2000). Per necessita famigliare: hypocrisy and corruption in fascist Italy. *Eur. Hist. Q.*, *30*(3), 357–387.

Bowen, P., Akintoye, A., Pearl, R., & Edwards, P. J. (2007). Ethical behaviour in the south African construction industry. *Constr. Manag. Econ.*, *25*(6), 631–648.

Bowen, P., Edwards, P., & Cattell, K. (2015). Corruption in the south African construction industry: experiences of clients and construction professionals. *Int. J. Proj. Organ. Manag.*, *7*(1), 72.

Bradshaw, J. M. (1997). *Software agents*. Cambridge: MIT Press.

Bremer, W., & Kok, K. (2000). The Dutch construction industry: a combination of competition and corporatism. *Build. Res. Inform.*, *28*(2), 98–108.

Brigham, E. F., & Ehrhardt, M. C. (2013). *Financial management: Theory & practice*. USA: Cengage Learning. Boston, MA.

Broadbent, J., & Laughlin, R. (1999). The private finance initiative: clarification of a future research agenda. *Finan. Account. Manag.*, *15*(2), 95–114.

Brookes, N. J., Hickey, R., Littau, P., Locatelli, G., & Oliomogbe, G. (2015). Using multi-case approaches in project management research: learning from the MEGAPROJECT experience. In: Pasian, B. (Ed.), *Designs, Methods and Practices for Research of Project Management*. New York, USA. Routledge. 1–12.

Brookes, N. J., & Locatelli, G. (2015). Power plants asmegaprojects: using empirics to shape policy, planning, and construction management. *Util. Policy*, *36*, 57–66.

Brown, K. (2007). Are public–private transactions the future of infrastructure finance? *Public Works Manag. Policy*, *12*(1), 320–324.

Bryan, C., Adams, G., & Monin, B. (2012). When cheating would make you a cheater: implicating the self prevents unethical behaviour. *J. Exp. Psychol. General*, *142*, 1–25. http://dx.doi.org/10.1037/a0030655

Buckner, R., & Carroll, D. (2007). Self-projection and the brain. *Trends Cognitive Sci.*, *11*, 49–57.

Cabrera, M., Suarez-Alemán, A., & Trujillo, L. (2015) Public-private partnerships in Spanish ports: current status and future prospects. *Utilities Policy*, *32*, 1–11.

Calandri, M. (2004, October 10), *Il pm chiede il processo per i vip dell' Alta Velocità*, Repubblica.

Camera dei deputati (1964). *Domanda di autorizzazione a procedere contro i deputati de Leonardis*, de Marzio, Ferri, Giglia, Mazzoni, Sangalli, Scarascia, Belotti, Vicentini. available at: http://legislature.camera.it/_dati/leg03/lavori/stampati/pdf/002_239001.pdf.

Camera dei deputati (2015a). Elenco interventi: Lombardia, Sistema Informativo Legge Opere Strategiche. available at: http://silos.infrastrutturestrategiche.it/opere/opere.aspx#moveHere (accessed 23 September 2015).

Camera dei deputati (2015b). Elenco Interventi: Piemonte. Sistema Informativo Legge Opere Strategiche (available at: http://silos.infrastrutturestrategiche. it/opere/opere.aspx?id=1 accessed 23 September 2015).

Campen, R (2009) Unpublished presentation to the International Working Group of Koninklijk Nederlands Waternetwerk, Deventer, 12 February 2009.

Cantarelli, C., & Flyvbjerg, B. (2015). Decision making and major transport infrastructure projects: the role of project ownership. In: Hickman, R., Givoni, M., Bonilla, D., & Banister, D. (Eds.), *Handbook on Transport and Development* (pp. 380–393). Cheltenham: Edward Elgar.

Cantarelli, C. C., Flyvbjerg, B., Molin, E. J. E., & van Wee, B. (2010). Cost overruns in large-scale transportation infrastructure projects: explanations and their theoretical embeddedness. *Eur. J. Transp. Infrastruct. Res.*, *10*(1), 5–18.

Cantarelli, C. C., Van Wee, B., Molin, E. J. E., & Flyvbjerg, B. (2012). Different cost performance: different determinants? The case of cost overruns in Dutch transport infrastructure projects. *Transp. Policy*, *22*, 88–95.

Carbonara, N., Costantino, N., & Pellegrino, R. (2014). Concession period for PPPs: a win–win model for a fair risk sharing. *Int. J. Project Manag.*, *32*(7), 1223–1232.

Carrasco, I. J., & Chang, S. Y. (2005). Random Monte Carlo simulation analysis and risk assessment for ammonia concentrations in wastewater effluent disposal. *Stochastic Environ. Rese. Risk Assess.*, *19*(2), 134–145.

Cartwright, E. (2011). *Behavioural economics*. s.l.: Taylor & Francis.

CGIA (2014). Grandi opere: costi e tempi infiniti. available at http://www.lent epubblica.it/wp-content/uploads/2015/03/Opere-strategiche.pdf.

Chai, J., Liu, J. N., & Ngai, E. W. (2013). Application of decision-making techniques in supplier selection: a systematic review of literature. *Exp. Syst. Appl.*, *40*(10), 3872–3885.

Chan, A., Yeung, J., Yu, C., Wang, S. Q., & Ke, Y. (2011). Empirical study of risk assessment and allocation of public-private partnership projects in China. *J. Manag. Eng.*, *27*(3), 136–148.

Chan, A. P., Lam, P. T., Chan, D. W., Cheung, E., & Ke, Y. (2010). Critical success factors for PPPs in infrastructure developments: chinese perspective. *J. Constr. Eng. Manag.*, *136*(5), 484–494.

Chan, A. P., Yeung, J. F., Yu, C. C., Wang, S. Q., & Ke, Y. (2011). Empirical study of risk assessment and allocation of public-private partnership projects in China. *J. Manag. Eng.*, *27*(3), 136–148.

Chapman, C., & Ward, S. (1996). *Project risk management: processes, techniques, and insights*. USA: John Wiley. Hoboken, New Jersey.

Chen, C. T., Lin, C. T., & Huang, S. F. (2006). A fuzzy approach for supplier evaluation and selection in supply chain management. *Int. J. Prod. Econ.*, *102*(2), 289–301.

Chiles, T. H., & McMackin, J. F. (1996). Integrating variable risk preferences trust and transaction cost economics. *Acad. Manag. Rev.*, *21*(1), 73–99.

Chinyio, E., & Gameson, R. (2009). Private finance initiative in use. In: Akintoye, A., & Beck, M. (eds) *Policy, finance & management for Public-Private Partnerships* (pp. 3–23). Oxford, UK: Wiley Blackwell.

Cho, I. & Kreps, D. (1987). Signaling games and stable equilibria. *Q. J. Econo.*, *102*, 179–221.

Chou, J. S. (2012). Comparison of multilabel classification models to forecast project dispute resolutions. *Expert Syst. Appl.*, *39*(11), 10202–10211.

Chou, J. S., & Leatemia, G. T. (2016). Critical process and factors for ex-post evaluation of public-private partnership infrastructure projects in Indonesia. *J. Manag. Eng.*, *32*(5), 05016011.

Chung, D., Hensher, D. A., & Rose, J. M. (2010). Toward the betterment of risk allocation: investigating risk perceptions of Australian stakeholder groups to public-private-partnership tollroad projects. *Res. Transp. Econ.*, *30*(1), 43–58.

Cialdini, R. B. (2006). *Influence: the psychology of persuasion*, revised edition. New York: William Morrow.

Cialdini, R. B., Reno, R. R., & Kallgren, C. (1990). A focus theory of normative conduct: recycling the concept of norms to reduce littering in public places. *J. Personality Social Psychol.*, *58*(6), 1015–1026. http://dx.doi.org/10.1037/0022-3514.58.6.1015

Cicconi, I. (2011). *Il Libro Nero dell'alta velocità*. KOINé.

Cicconi, I., De Benedetti, A., Giorno, C., & Paola, F. (2015). *TAV CHI SI, Fert rights.* available at: http://www.trancemedia.eu/screen-studio/nomedia/cart/add/content/TAV+CHI+SI#Inizio.

Cirillo, E. (1994, April). Alta velocità, oggi al via, Repubblica, Italy.

Clinard, M. B., & Quinney, R. (1973). *Criminal behaviour systems: a typology* (2nd ed.). New York, NY: Holt, Rinehart, and Winstron.

Copeland, T., & Antikarov, V. (2001). *Real Options Texere LLC Publishing, New YorkCopeland*, T., Antikarov, V., 2001. New York: Real Options Texere LLC Publishing.

Corrente, S., Figueira, J. R., & Greco, S. (2014). The smaa-promethee method. *Eur. J. Operat. Res.*, *239*(2), 514–522.

Corriere della sera. (1996, September 17), Riesplode Tangentopoli, arrestato Necci, *Corriere della sera.*

Corriere della sera. (2001, October 5), Pacini condannato per corruzione I suoi avvocati: prove inutilizzabili, *Corriere della sera.*

Corriere della sera. (2007, January 12), 'Tangentopoli due'. Molti reati prescritti, *Corriere della sera.*

Coulson, A. (1997). Trust and contract in public sector management. University of Birmingham Occasional Paper.

Coulson, A. (1998). Client–contractor relationships in ten local authorities. University of Birmingham Occasional Paper 18, 104.

Cox, J. C., Ross, S. A., & Rubinstein, M. (1979). Option pricing: a simplified approach. *J. Financ. Econ.*, *7*(3), 229–263.

Cruz C., & Marques R. (2013a) *Infrastructure public-private partnerships: decision, management and development.* London: Springer.

Cruz C., & Marques R. (2013b) Flexible contracts to cope with uncertainty in public-private partnerships. *Int. J. Project. Manag.*, *31*, 473–483.

Cuttaree, V. (2008). *Successes and failures of PPP projects.* Washington, DC: World Bank.

da Silva Pereira, E. J., Pinho, J. T., Galhardo, M. A. B., & Macêdo, W. N. (2014). Methodology of risk analysis by Monte Carlo Method applied to power generation with renewable energy. *Renewable Energy, 69,* 347–355.

Darley, J. M. (2005). The cognitive and social psychology of contagious organizational corruption. *Brooklyn Law Review, 70,* 1177–1194.

Darroch, F. (2007). Lesotho Highlands Water Project: corporate pressure on the prosecution and judiciary. In: *Global Corruption Report 2007,* Berlin: Transparency International. Available at: www.transparency.org/publications/gcr

Davis, H., Walker, B., & Coulson, A. (1998). Trust and Contracts: Relationships in Local Government, Health and Public Services. University of Birmingham Occasional Paper.

Davis, J (2004) Corruption in public service delivery: experience from South Asia's water and sanitation sector, *World Development, 32*(1), 53–71.

De Jong, M., Mu, R., Stead, D., Ma, Y., & Xi, B. (2010). Introducing public-private partnerships for metropolitan subways in China: what is the evidence? *J. Transp. Geogr., 18,* 301–313.

De Oliveira, L. B., & Camponogara, E. (2010) Multi-agent model predictive control of signaling split in urban traffic networks. *Transp. Res. Part C Emerg. Technol., 18,* 120–139.

Deakin, N., & Walsh, K. (1996). The enabling state: the role of markets and contracts. *Public Admin., 74*(1), 33–48.

Deakin, S., Lane, C., & Wilkinson, F. (1997). *Contract law, trust relations and incentives for co-operation: a comparative study.* In: Deakin, S., & Michie, J. (Eds.), Contracts, co-operation, and competition studies in economics, management, and law (pp. 105–139). Oxford: Oxford University Press.

Deakin, S., & Michie, J. (1997). The theory and practice of contracting. In: Deakin, S., & Michie, J. (Eds.), *Contracts, co-operation, and competition* (pp. 121–125). Oxford: Oxford University Press.

Della Port, A. D., & Vannucci, A. (1999). *Corrupt exchanges: actors, resources and mechanisms of political corruption.* New York: Aldine.

del-Rey-Chamorro, F. M., Roy, R., Wegen, B. V., & Steele, A. (2003). A framework to create key performance indicators for knowledge management solutions. *J. Knowled. Manag., 7*(2), 46–62.

Di Marco, M., Baker, M. L., Daszak, P., De Barro, P., Eskew, E. A., Godde, C. M., ... & Karesh, W. B. (2020). Opinion: sustainable development must account for pandemic risk. *Proc. Natl. Acad. Sci., 117*(8), 3888–3892.

Dickson, G. W. (1966). An analysis of vendor selection systems and decisions. *J. Purchas., 2*(1), 5–17.

Dixit, A. K., & Pindyck, R. S. (1994). *Investment under uncertainty.* Princeton, NJ: Princeton University Press.

Domberger, S. (1998). *The contracting organisation: a strategic guide to outsourcing.* New York: Oxford University Press.

Domingues, S., & Zlatkovic, D. (2015) Renegotiating PPP contracts: reinforcing the 'P' in partnership. *Trans. Rev. Transnatl. Transdiscipl. J.,* 35(2), 204–225.

Dong, B., Dulleck, U., & Torgler, B. (2012). Conditional corruption. *J. Econ. Psychol.,* 33(3), 609–627. http://doi.org/10.1016/j.joep.2011.12.001

Dungan, J., Waytz, A., & Young, L. (2014). Corruption in the context of moral trade-offs. *J. Interdisciplinary Econ.,* 26(1–2), 97–118. http://doi.org/10.1177/0260107914540832

Durbach, I. (2006). A simulation-based test of stochastic multicriteria acceptability analysis using achievement functions. *Eur. J. Oper. Res.,* 170(3), 923–934.

Durbach, I. (2009). On the estimation of a satisficing model of choice using stochastic multicriteria acceptability analysis. *Omega,* 37(3), 497–509.

Durbach, I. N. (2009). The use of the SMAA acceptability index in descriptive decision analysis. *Eur. J. Oper. Res.,* 196(3), 1229–1237.

Durbach, I. N., & Calder, J. M. (2016). Modelling uncertainty in stochastic multicriteria acceptability analysis. *Omega,* 64, 13–23.

Earle, A., & Turton, A (2005). No duck no dinner: how sole sourcing triggered Lesotho's struggle against corruption. In: *African Water* Issues Research Unit (AWIRU) paper, Pretoria: University of Pretoria, CiPS. Available at: www.acwr.co.za/pdf_files/07.pdf

EC (European Commission). (2005). Guidelines for successful public-private partnerships, *EC Directorate General,* Regional Policy, March 2003.

Edwards, P., & Bowen, P. (2003). Risk perception and communication in public-private partnerships. *Public-Private Partnerships,* 79–92.

Egger, P., & Winner, H. (2005). Evidence on corruption as an incentive for foreign direct investment. *Eur. J. Polit. Econ.,* 21(4), 932–952.

EIB (2004). The EIB's role in public-private partnerships (PPPs). Brussels, Luxembourg: European Investment Bank.

EIB (2012). The guide to guidance: how to prepare, procure and deliver PPP projects. Brussels, Luxembourg: European Investment Bank.

El-Gohary, N. M., Osman, H., & El-Diraby, T. E. (2006). Stakeholder management for public private partnerships. *Int. J. Project Manag.,* 24(7), 595–604.

Eliasberg, W. (1951). Corruption and bribery. *J. Criminal Law, Criminology, Police Sci.,* 42, 317–331. URL: http://www.jstor.org/stable/1140346?seq=2

Ellis, J. (2019). *Corruption, social sciences, and the law: exploration across the disciplines.* New York, USA: Routledge.

Elster, J. (1989). *The cement of society: a survey of social order.* New York, NY: Cambridge University Press.

Elton, E., & Gruber, M. (1995). *Modern Portfolio Theory and Investment Analysis,* 5th ed. New York: Wiley.

Elyamany, A., Basha, I., & Zayed, T. (2007). Performance evaluation model for construction companies. *J. Constr. Eng. Manag.,* 133(8), 574–581.

Emilia-Romagna, R. (2015). Emilia Romagna mobilità: ferrovie. available at: http://mobilita.regione.emilia-romagna.it/ferrovie/sezioni/alta-velocita-2/latratta-bologna-firenze-1 (accessed 23 September 2015).

European Commission (2003). Guidelines for successful public-private partnerships. Brussels, Belgium: European Commission.

European Commission (2015). Infrastructure - TEN-T - connecting Europe. available at http://ec.europa.eu/transport/themes/infrastructure/ten-t-guideli nes/project-funding/cef_en.htm.

Farrell, S., & Vanelslander, T. (2015) Comparison of public-private partnerships in airports and seaports in low- and middleincome countries. *Trans. Rev. Transnatl. Transdiscipl. J.*, *35*(3): 329–351.

Fender, I., & Kiff, J. (2004). CDO rating methodology: some thoughts on model risk and its implications.

Fernández, A., Javier, F., & Atienza, V. J. (2012). High-Speed Line Costs Internalized in the Infrastructure.Manager's Accounts. *2*, Publication: 360.high speed magazine No 2 - May 2012, pp. 5–22.

Ferrovie dello Stato (2015). Le linee Alta Velocità: storia e traguardi. available at: http://www.fsitaliane.it/cms-file/allegati/il-gruppo/Linee_AVstoria_traguardi.pdf.

Festinger, L., & Carlsmith, J. (1959). Cognitive consequences of forced compliance. *The Journal of Abnormal and Social Psychology*, *58*, 203–210. doi:10.1037/h0041593

Finney, H. C., & Lesieur, H. R. (1982). A contingency theory of organizational crime. In: S. B. Bacharach (Ed.), *Research in the sociology of organizations* (Vol. *1*, pp. 255–299). Greenwich, CT: JAI Press.

Fitzgerald, P. (2004). Review of partnerships Victoria provided infrastructure. Final Report to the Treasurer. Melbourne, Australia: Growth Solutions Group.

Fleta-Asín, J., Muñoz, F., & Rosell-Martínez, J. (2019). Public-private partnerships: determinants of the type of governance structure. *Public Manag. Rev.*, *22*, 1–26.

Flyvbjerg, B. (2008). *Public Planning of Mega-Projects: Overestimation of Demand and Underestimation of Costs.* In: Flyvbjerg, B., Priemus, H., & van Wee, B. (Eds.), Decision-making on mega-projects. Cost–benefit analysis, planning and innovation (pp. 120–144). Cheltenham, UK: Edward Elgar Publishing.

Flyvbjerg, B., Bruzelius, N., & Rothengatter, W. (2003). *Megaprojects and Risk, Megaprojects and Risk: An Anatomy of Ambition.* Cambridge: Cambridge University Press.

Flyvbjerg, B., Hon, C., & For, W. H. (2016). Reference Class Forecasting for Hong Kong's Major Roadworks Projects. *Proc. Inst. Civil Eng.-Civil Eng.*, *169*(6), 17–24.

Flyvbjerg, B., & Molloy, E. (2011). Delusion, deception and corruption in major infrastructure projects: causes, consequences and cures. *International handbook on the economics of corruption* Vol. Two. Edward Elgar Publishing Ltd., pp. 81–107. Cheltenham, United Kingdom.

Froud, J., & Shaoul, J. (2001). Appraising and evaluating PFI for NHS Hospitals. *Financial Accountability & Management*, *17*(3), 247–270.

Fudenberg, D., & Tirole, J. (1991). *Game Theory.* Cambridge, MA: The MIT Press.

Furmston, M. P., Cheshire, G. C., & Fifoot, C. H. S. (2012). *Cheshire, Fifoot and Furmston's Law of Contract.* UK: Oxford University Press, Oxford, United Kingdom.

Gaffney, D., Pollock, A. M., Price, D., & Shaoul, J., 1999. The politics of the private finance initiative and the new NHS. *British Med. J.*, *319*(7204), 249–253. https://doi.org/10.1136/bmj.319.7204.249

Garemo, N., Matzinger, S., & Palter, R. (2015). Megaprojects: the good, the bad, and the better. available at: http://www.mckinsey.com/insights/infrastructure/megaprojects_the_good_the_bad_and_the_better.

Garvin, M., Molenaar, K., Navarro, D., & Proctor, G. (2011). Key performance indicators in Public-Private Partnerships. Report to U.S. Department of Transportation and Federal Highway Administration, FHWA-PL-10-029, American Trade Initiatives, Alexandria, USA.

Ghodsypour, S. H., & O'brien, C. (2001). The total cost of logistics in supplier selection, under conditions of multiple sourcing, multiple criteria, and capacity constraint. *Int. J. Production Econ.*, *73*(1), 15–27.

Ghosal, S., & Moran, P. (1996). Bad for practice: a critique of transaction cost theory. *Academy of Management Review 21*(1), 13–47.

GIACC (2014). What is corruption, Global Infrastructure Anti-Corruption Centre, available at: http://www.giaccentre.org/what_is_corruption.php.

GIACC. (2008), Examples of corruption in infrastructure, (available at:) http://www.giaccentre.org/documents/GIACC.CORRUPTIONEXAMPLES.pdf.

Gibbons, R. (1992). *Game Theory for Applied Economists*. Princeton, NJ: Princeton University Press.

Giezen, M. (2012). Keeping it simple? A case study into the advantages and disadvantages of reducing complexity in mega project planning. *Int. J. Proj. Manag.* *30*(7), 781–790.

Gigerenzer, G. & Selten, R. (2002). Bounded rationality: the adaptive toolbox. Dahlem Workshop Reports ed. Cambridge, MA: The MIT Press.

Gilbert, D. (2006). Stumbling on Happiness. New York, NY: Knopf.

Gilbert, D., & Wilson, T. (2007). Prospection: experiencing the future. *Science*, *317*(September), 1351–1355. http://doi.org/10.1126/science.1144161

Gillanders, R. (2014). Corruption and infrastructure at the country and regional level. *J. Dev. Stud. 50*(6), 803–819.

Gino, F., & Bazerman, M. H. (2009). When misconduct goes unnoticed: the acceptability of gradual erosion in others' unethical behavior. *J. Exp. Social Psychol.*, *45*, 708–719. doi: 10.1016/j.jesp.2009.03.013

González de Asís, M, O'Leary, D, Ljung, P and Butterworth, J (2009). *Improving transparency, accountability, and integrity in water supply and sanitation: action, learning, experiences*, Washington, DC: World Bank.

Graf Lambsdorff, J., Taube, M., & Schramm, M. (2004). *The New Institutional Economics of Corruption*. New York, USA: Routledge.

Graycar, A., & Smith, R. G. (2011). *Handbook of global research and practice in corruption*. Northampton, MA: Edward Elgar.

Grimsey, D., & Lewis, M. K. (2002). Evaluating the risks of public private partnerships for infrastructure projects. *Int. J. Proj. Manag.*, *20*(2), 107–118.

Grimsey, D., & Lewis, M. K. (2005). Are public private partnerships value for money? *Accounting Forum*, *29*(4), 345–378.

Grimsey, D., & Lewis, M. (2007). *Public private partnerships: The worldwide revolution in infrastructure provision and project finance*. Cheltenham, UK: Edward Elgar Publishing.

Grimsey, D., & Lewis, M. K. (2002). Evaluating the risks of public private partnerships for infrastructure projects. *Int. J. Proj. Manag.*, *20*(2), 107–118.

Grout, P. A. (1997). The economics of the private finance initiative. *Oxford Rev. Econ. Policy*, *13*(4), 53–66.

Haidt, J., & Kesebir, S. (2010). Morality. In: S. Fiske, D. Gilbert, & G. Lindzey (Eds.), *Handbook of Social Psychology* (5th ed., pp. 797–832). New York, New York University.

Hall, D. (1999). Privatisation, multinationals, and corruption. *Develop. Practice, 9*(5), 539–556.

Han, H., & Shum, M. (2002). Increasing competition and the winner's curse: evidence from procurement. *Rev. Econo. Stud., 69*, 871–898.

Han, Z. (2013). *Procedures and resources for analyses related to public-private partnerships* (Doctoral dissertation).

Han, Z., Porras-Alvarado, J. D., Sun, J., & Zhang, Z. (2017). Monte Carlo simulation–based assessment of risks associated with public–private partnership investments in toll highway infrastructure. *Transp. Res. Rec., 2670*(1), 59–67.

Hargreaves Heap, S. P., & Varoufakis, Y. (2004). *Game theory: a critical text*, second ed. London and New York: Routledge.

Hart, O. (2003). Incomplete contracts and public ownership: remarks, and an application to public-private partnerships. *Econ J, 113*(486), 69–76.

Haskins, S., Gale, D., & Kelly, L. (2002). Creating and optimizing new forms of public-private partnerships in Seattle. *Water Science Technology: Water Supply, 2*(4), 211–218.

Hauser, C., & Hogenacker, J. (2014). Do firms proactively take measures to prevent corruption in their international operations? *Eur. Manag. Rev., 11*(3–4), 223–237.

Heidenheimer, A. J. (1970). *Political corruption: readings in comparative analysis.* New York, New Brunswick: Transaction Publishers.

Heidenheimer, A. J., Johnson, M., & LeVine, V. T. (1989). *Political corruption: a handbook*Oxford, UK: Transaction Publishers.

Hellman, J. S., Jones, G., Kaufmann, D., & Schankerman, M. (2000). Measuring governance, corruption, and state capture: how firms and bureaucrats shape the business environment in transition economies (World Bank Policy Research Working Paper No. 3212). Washington DC: World Bank Publishing Group.

Henisz, W. J., Levitt, R. E., & Scott, W. R. (2012). Toward a unified theory of project governance: economic, sociological and psychological supports for relational contracting. *Eng. Proj. Organ. J., 2*(1–2), 37–55.

Henjewele, C., Sun, M., & Fewings, P. (2011). Critical parameters influencing value for money variations in PFI projects in the healthcare and transport sectors. *Constr. Manag. Econo., 29*(8), 825–839.

Henrich, J., Ensminger, J., McElreath, R., Barr, A., Barrett, C., Bolyanatz, A., … Ziker, J. (2010). Markets, religion, community size, and the evolution of fairness and punishment. *Science, 327*(5972), 1480–1484. http://doi.org/10.1126/science.1182238

Heravi, G., & Hajihosseini, Z. (2012). Risk allocation in public–private partnership infrastructure projects in developing countries: case study of the Tehran–Chalus toll road. *J. Infrastruct. Syst., 18*(3), 210–217.

Heritage Foundation (2012). *Index of economic free*dom, Washington D.C.: HF.

Heydari, M (2021). *A cognitive basis perceived corruption and attitudes towards entrepreneurial intention*, PhD Thesis. Nanjing, Jiangsu, China: Nanjing University of Science and Technology,.

Heydari, M., Lai, K. K., & Xiaohu, Z. (2020). *Risk management in public-private partnerships.* Taylor & Francis Group. New York & UK, London.

HM Treasury (2003). *PFI: meeting the investment challenge*. London: HMSO.

Ho, S. P. (2001). *Real options and game theoretic valuation, financing and tendering for investments on build-operate-transfer projects*. Ph.D. Thesis. Urbana, IL: Department of Civil and Environmental Engineering, University of Illinois at Urbana-Champaign.

Ho, S. P. (2001). *Real options and game theoretic valuation, financing, and tendering for investments on build-operate-transfer projects* (Doctoral dissertation, the University of Illinois at Urbana-Champaign).

Ho, S. P. (2005). Bid compensation decision model for projects with costly bid preparation. *J. Constr. Eng. Manag.*, *131*(2), 151–159. 7 (in Chinese).

Ho, S. P. (2006). Model for financial renegotiation in public-private partnership projects and its policy implications: game theoretic view. *J. Constr. Eng. Manag.*, *132*(7), 678–688.

Ho, S. P. & Liu, L. Y. (2004). Analytical model for analyzing construction claims and opportunistic bidding. *J. Constr. Eng. Manag.*, *130*(1), 94–104.

Ho, S. P. & Tsui, C. (2009). The transaction costs of public-private partnerships: implications on PPP governance design. Lead 2009 Specialty Conference: Global Governance in Project Organizations, South Lake Tahoe, CA.

Ho, S. P. & Tsui, C. (2010). When are public-private partnerships not an appropriate governance structure? Case study evidence. *Proceedings of Construction Research Congress 2010*, Banff, AB.

Ho, S. P., & Tsui, C. W. (2009, November). The transaction costs of Public-Private Partnerships: implications on PPP governance design. In *Lead 2009 Specialty Conference: Global Governance in Project Organizations*, South Lake Tahoe, CA (pp. 5–7).

Ho, S. P., & Tsui, C. W. (2010). When are public-private partnerships, not an appropriate governance structure? Case study evidence. In the *Construction Research Congress 2010: Innovation for Reshaping Construction Practice* (pp. 817–826).

Ho, W., Xu, X., & Dey, P. K. (2010). Multi-criteria decision-making approaches for supplier evaluation and selection: a literature review. *Eur. J. Oper. Res.*, *202*(1), 16–24.

Hodge, G. A. (2004). The risky business of public-private partnerships. *Austr. J. Public Admin.*, *63*(4), 37–49.

Hodge, G. A. (2005). Public private partnerships: the Australasian experience with physical infrastructure. In: Hodge, G. A. and Greve, C. (eds) *The challenge of public-private partnerships: learning from international experience* (pp. 305–331). Cheltenham, UK: Edward Elgar.

Hogg, M. A., Hohman, Z. P., & Rivera, J. E. (2008). Why do people join groups? Three motivational accounts from social psychology. *Social Personality Psychol. Compass*, *2*(3), 1269–1280. http://doi.org/10.1111/j.1751-9004.2008.00099.x

Holliday, I., Marcou, G., & Vickerman, R. (1991). *The Channel Tunnel: public policy, regional development and European integration*. New York: Belhaven Press.

House of Commons (2003). The operational performance of PFI prisons. London, UK:·House of Commons.

Huang, Z., Zheng, P., Ma, Y., Li, X., Xu, W., & Zhu, W. (2016). A simulation study of the impact of the public–private partnership strategy on the performance of transport infrastructure. *SpringerPlus*, 5(1), 1–15.

Hwang, B. G., Zhao, X., & Yu, G. S. (2016). Risk identification and allocation in underground rail construction joint ventures: contractors' perspective. *J. Civil Eng. Manag.*, 22(6), 758–767.

Il fatto quotidiano. (2014, December 15), Corruzione, Cantone: 'Dopo Tangentopoli nessunmeccanismo per arginarla', Il fatto quotidiano, available at: http://www.ilfattoquotidiano.it/2014/12/15/corruzione-cantone-dopo-tangentopoli-n essunmeccanismo-per-arginarla/1275412/.

Il fatto quotidiano. (2015, March16), Ercole Incalza, in carcere superburocrateLavori pubblici. 'Corruzione in Tav', Il fatto quotidiano, available at: http://www.ilfa ttoquotidiano.it/2015/03/16/ercole-incalza-in-carcere-dirigentedei-lavori-p ubblici-inchiesta-procura-firenze/1508213/.

Infrastructure Partnerships Australia (2007). *Performance of PPPs and traditional procurement in Australia*. Sydney 2007, 21–26.

IPMA (2015). Code of ethics and professional conduct, available at: http://www.ipma.world/assets/IPMA-Code-of-Ethics-and-Professional-Conduct.pdf.

Irwin, T., & Mokdad, T. (2010). *Managing contingent liabilities in public-private partnerships: Practice in Australia, Chile, and South Africa*. World Bank.

Ismail, S. (2013). Critical success factors of public private partnership (PPP) implementation in Malaysia. *Asia-Pac. J. Bus. Admin*, 5(1), 6–19.

Italo (2015). Acquisto bilglietto. available at: http://www.italotreno.it/.

Jensen, M. C., & Smith, C. W. J. (2000). Stockholder, manager, and creditor interests: applications of agency theory. Theory of the firm: governance, residual claims, and organizational forms. Cambridge, MA: Harvard University Press.

Jafarizadeh, B., & Khorshid-Doust, R. R. (2008). A method of project selection based on capital asset pricing theories in a framework of mean–semideviation behavior. *Int. J. Proj. Manag.*, 26(6), 612–619.

Javid, M., & Seneviratne, P. N. (2000). Investment risk analysis in airport parking facility development. *J. Constr. Eng. Manag.*, 126(4), 298–305.

Jeon, C. M., & Amekudzi, A. A. (2007). *Risk Management for Public-Private Partnerships: Using Scenario Planning and Valuation Methods* (07–0793). Transportation Research Board 86th Annual Meeting, January 21, 2007 to January 21, 2007. Washington, DC.

Jones, T. M. (1991). Ethical decision-making by individuals in organizations - an issue-contingent model. *Acad. Manage. Rev.*, 16(2), 366–395. http://doi.org/10.5465/AMR.1991.4278958

Judge, W. Q., McNatt, D. B. & Xu, W. (2011). The antecedents and effects of national corruption: a meta-analysis. *J. World Business*, 46(01), 93–103. URL: http://www.sciencedirect.com/science/article/pii/S1090951610000362

Kagioglou, M., Cooper, R., & Aouad, G. (2001). Performance management in construction: a conceptual model. *Constr. Manag. Econo.*, 19(1), 85–95.

Kakimoto, R., & Seneviratne, P. N. (2000). Investment risk analysis in port infrastructure appraisal. *J. Infrastruct. Syst.*, 6(4), 123–129.

Kalidindi, S. N., & Thomas, A. V. (2003). Private sector participation road projects in India: assessment and allocation of critical risks. In: Akintoye, A., Beck, M., &

Hardcastle, C. (eds.) *Public-Private Partnerships*, Hoboken, NJ: Wiley, p. 317. https://doi.org/10.1002/9780470690703.ch15

Kangas, A. S., Kangas, J., Lahdelma, R., & Salminen, P. (2006). Using SMAA-2 method with dependent uncertainties for strategic forest planning. *For. Policy Econ.*, *9*(2), 113–125.

Karklins, R. (2005). The system made me do it: corruption in post-communist societies. Armonk, New York, NY: M.E. Sharpe.

Kaufmann, D., & Wei, S.-J. (1999). *Does "Grease Money" Speed up the Wheels of Commerce?* NBER Working Paper (available at: https://www.imf.org/external/pubs/ft/wp/2000/wp0064.pdf.)

Kavanagh, I., & Parker, D. (2000). Contracting out of local government services in the UK: a study in transaction costs. Aston Business School Occasional Paper.

Kenny, C. (2006) Construction, Corruption, and Developing Countries. In: World Bank Policy Research Working Paper No. 4271. Washington, DC: World Bank.

Kenny, C. (2007) Infrastructure governance and corruption: where next?. In: World Bank Policy Research Working Paper 4331. Washington, DC: World Bank.

Kenny, C. (2006). Measuring and reducing the impact of corruption in infrastructure. World Bank Policy Research Working Paper (available at: http://elibrary.worldbank.org/doi/pdf/10.1596/1813-9450-4099).

Kenny, C. (2012). Publishing Construction Contracts to Improve Efficiency and Governance. *Proc. ICE - Civil Eng.*, *165*(5), 18–22.

Kenny, C., Klein, M. U., Sztajerowska, M. (2011). A trio of perspectives on corruption: bias, speed money and "grand theft infrastructure. World Bank Policy Research Working Paper Vol. No. 5889. http://dx.doi.org/10.2139/ssrn.1965920.

Keown, A. J., Martin, J. D., Petty, J. W., & Scott Jr, D. F. (2005). Financial management-principles and applications Prentice Hall. Tenth International Edition.

Kerr, N. L., & Kaufman-Gilliland, C. M. (1997). "... and besides, I probably could not have made a difference anyway": Justification of Social Dilemma Defection via Perceived Self-Inefficacy. *J. Exp. Social Psychol.*, *33*(3), 211–230. http://doi.org/10.1006/jesp.1996.1319

Khan, M. Y., & Jain, P. K. (2005). *Basic Financial Management*. USA: Tata McGraw-Hill Education Pvt. Ltd. New York, NY.

Kipnis, D., Castell, P. J., Gergen, M., & Mauch, D. (1976). Metamorphic effects of power. *J. Appl. Psychol.*, *61*, 127–135. doi:10.1037/0021-9010.61.2.127

Klemperer, P. (2004). *Auctions: theory and practice*. Princeton and Oxford: Princeton University Press.

Klitgaard, R. (1988). *Controlling corruption*. Berkeley, CA: University of California Press.

Ko, K. & Samajdar, A. (2010). Evaluation of international corruption indexes: should we believe them or not? *Social Sci. J.*, *47*, 508–540. http://www.sciencedirect.com/science/article/pii/S0362331910000352

Köbis, N. C., van Prooijen, J.-W., Righetti, F., & Van Lange, P. A. M. (2015). "Who doesn't?"- The impact of descriptive norms on corruption. *PloS One, 10*, e0131830. doi: 10.1371/journal.pone.0131830

Köbis, N. C., van Prooijen, J.-W., Righetti, F., & Van Lange, P. A. M. (2016). Prospection in individual and interpersonal corruption dilemmas. *Rev. General Psychol.*, *20*(1), 71–85.

Köbis, N. C. (2018), 'The social psychology of corruption,' Ph.D., Vrije Universiteit Amsterdam.

Kosfeld, M. (1997). *Corruption within a Cooperative Society* No 48, Economics Series, Institute for Advanced Studies, https://EconPapers.repec.org/RePEc:ihs:ihsesp:48.

Krüger, N. (2012) To kill a real option—incomplete contracts, real options and PPP. *Transp Res Part A*, *46*, 1359–1371.

Kurer, O. (2005). Corruption: an alternative approach to its definition and measurement (?): A review. *Political Studies*, *53*, 222–239. https://doi.org/10.1111/j.1467-9248.2005.00525.x

Kwak, Y. H., Chih, Y. Y., & Ibbs, C. W. (2009). Towards a comprehensive understanding of public private partnerships for infrastructure development. *California Manag. Rev.*, *51*(2), 51–78.

KWTX (2005). *Farmers and Ranchers Rally Against Trans Texas Corridor Project.* London, UK: Sage.

KXII. Trans Texas Corridor—Misconceptions (2006). http://www.kxii.com/home/headlines/3381786.html. Accessed July 28, 2016.

Lahdelma, R., & Salminen, P. (2001). SMAA-2: Stochastic multicriteria acceptability analysis for group decision making. *Oper. Res.*, *49*(3), 444–454.

Lahdelma, R., & Salminen, P. (2006). Classifying efficient alternatives in SMAA using cross confidence factors. *Eur. J. Oper. Res.*, *170*(1), 228–240.

Lahdelma, R., & Salminen, P. (2006). Stochastic multicriteria acceptability analysis using the data envelopment model. *Eur. J. Oper. Res.*, *170*(1), 241–252.

Lahdelma, R., & Salminen, P. (2009). Prospect theory and stochastic multicriteria acceptability analysis (SMAA). *Omega*, *37*(5), 961–971.

Lahdelma, R., Hokkanen, J., & Salminen, P. (1998). SMAA-stochastic multiobjective acceptability analysis. *Eur. J. Oper. Res.*, *106*(1), 137–143.

Lahdelma, R., Salminen, P., & Hokkanen, J. (2002). Locating a waste treatment facility by using stochastic multicriteria acceptability analysis with ordinal criteria. *Eur. J. Oper. Res.*, *142*(2), 345–356.

Lambsdorff, J. G. (1999). The Transparency International corruption perceptions index 1999: Framework document. Berlin: Transparency International. *www.transparency. De [13.12. 00].*

Lambsdorff, J. G., & Frank, B. (2011). Corrupt reciprocity – experimental evidence on a men's game. *Int. Rev. Law Econ.*, *31*(2), 116–125. http://doi.org/10.1016/j.irle.2011.04.002

Lambsdorff, J. G. (2012). Behavioural and experimental economics as guidance to anticorruption. In: D. Serra & L. Wantchekon, eds. *New advances in experimental research on corruption research in experimental economics* (pp. 279–299). 15 ed. s.l.: Emerald Group Publishing Limited. URL: http://www.emeraldinsight.com/books.htm?chapterid=17033202

Lambsdorff, J. G., Taube, M., & Schramm, M. (2005). *The new institutional economics of corruption.* New York, NY: Routledge.

Lambsdorff, J. G. (2003). How corruption affects productivity. *Kyklos, 56*(4), 457–474.

Lane, J.-E. (1999). Contractualism in the public sector: some theoretical considerations. *Public Manag., 1*(2), 179–193.

Lee, H. (2011). *A Real option approach to valuating infrastructure investments* (Doctoral dissertation, KDI School).

Lee, J.-K. (2008). Cost overrun and cause in Korean social overhead capital projects: roads, rails, airports, and ports. *J. Urban Plann. Dev., 134*(2), 59–62.

Lee-Chai, A. Y., & Bargh, J. (2001). The use and abuse of power: multiple perspectives on the causes of corruption. Philadelphia, PA: Psychology Press.

Leff, N. H. (1964). Economic development through bureaucratic corruption. *Am. Behav. Sci., 8*(3), 8–14.

Leiringer, R. (2006). Technical innovation in PPPs: incentives, opportunities and actions. *Constr. Manag. Econo., 24*(3), 301–308.

Leonard, D., & McAdam, R. (2002). The role of the business excellence model in operational and strategic decision making. *Manag. Decision, 40*(1), 17–25.

Li, B., Akintoye, A., Edwards, P. J., & Hardcastle, C. (2005). The allocation of risk in PPP/PFI construction projects in the UK. *Int. J. Proj. Manag., 23*(1): 25–35.

Liang, R., Wu, C., Sheng, Z., & Wang, X. (2018). Multi-criterion two-sided matching of public–private partnership infrastructure projects: criteria and methods. *Sustainability, 10*(4), 1178.

Liautaud, G. (2004). Maintaining roads: experience with output-based contracts in Argentina. Washington, D.C: The World Bank.

Lien, D.-H. D. (1988). Coalitions in competitive bribery games. *Math. Soc. Sci., 15*(2), 189–196.

Ling, F. Y. Y., Ong, S. Y., Ke, Y., Wang, S., & Zou, P. (2014). Drivers and barriers to adopting relational contracting practices in public projects: comparative study of Beijing and Sydney. *Int. J. Proj. Manag., 32*(2), 275–285.

Ling, F. Y. Y., & Tran, P. Q. (2012). Effects of interpersonal relations on public sector construction contracts in Vietnam. *Constr. Manag. Econ., 30*(12), 1–15.

Liu, J., Love, P. E. D., Carey, B., Smith, J., & Regan, M. (2014a). Ex-ante evaluation of public-private partnerships: macroeconomic analysis. *J. Infrastruct. Syst.,* 10.1061/(ASCE)IS.1943-555X.0000228, 04014038.

Liu, J., Love, P. E. D., Smith, J., Regan, M., & Davis, P. R. (2014c). Life cycle critical success factors for public-private partnership infrastructure projects. *J. Manag. Eng.,* 10.1061/(ASCE)ME.1943-5479.0000307, 04014073.

Liu, J., Love, P. E. D., Smith, J., Regan, M., & Sutrisna, M. (2014d). Public-private partnerships: a review of theory and practice of performance measurement. *Int. J. Prod. Perform. Manag., 63*(4), 499–512.

Liu, Love, P. E. D., Davis, P. R., Smith, J., & Regan, M. (2014b). Conceptual framework for the performance measurement of public-private partnerships. *J. Infrastruct. Syst.,* 10.1061/(ASCE)IS.1943-555X.0000210, 04014023.

Liu, J., Love, P. E., Davis, P. R., Smith, J., & Regan, M. (2015). Conceptual framework for the performance measurement of public-private partnerships. *J. Infrastruct. Syst., 21*(1), 04014023.

Liu, Q., Liao, Z., Guo, Q., Degefu, D. M., Wang, S., & Jian, F. (2019). Effects of short-term uncertainties on the revenue estimation of PPP sewage treatment projects. *Water, 11*(6), 1203.

Liu, T., Wang, Y., & Wilkinson, S. (2016). Identifying critical factors affecting the effectiveness and efficiency of tendering processes in Public–Private Partnerships (PPPs): a comparative analysis of Australia and China. *Int. J. Proj. Manag.*, *34*(4), 701–716.

Liyanage, C., & Villalba-Romero, F. (2015). Measuring success of PPP transport projects: a cross-case analysis of toll roads. *Trans. Rev. Transnatl. Transdiscipl. J.*, *35*(2), 140–161.

Locatelli, G., Mancini, M., & Romano, E. (2014). Systems engineering to improve the governance in complex project environments. *Int. J. Proj. Manag.*, *32*(8), 1395–1410.

Locatelli, G., Mariani, G., Sainati, T., & Greco, M. (2017). Corruption in public projects and megaprojects: there is an elephant in the room!. *Int. J. Proj. Manag.*, *35*(3), 252–268.

Loosemore, M., & Lim, B. (2015). Inter-organizational unfairness in the construction industry. *Constr. Manag. Econ.*, *33*(4), 310–326.

Love, P. E. D., & Holt, G. (2000). Construction business performance measurement: the SPM alternative. *Business Process Manag. J.*, *6*(5), 408–416.

Lui, F. T. (1985). An equilibrium queuing model of bribery. *J. Polit. Econ.*, *93*(4), 760.

Ma, C., & Xu, C. (2009). Study on general and sectoral characteristics of the corruption in project bidding. In: 2009 16th International Conference on Industrial Engineering and Engineering Management. Beijing, China: IEEE, pp. 274–278.

Ma, H., Zeng, S., Lin, H., & Zeng, R. (2020). Impact of public sector on sustainability of public–private partnership projects. *J. Constr. Eng. Manag.*, *146*(2), 04019104.

Ma, J., Fan, Z. P., & Huang, L. H. (1999). A subjective and objective integrated approach to determine attribute weights. *Eur. J. Oper. Res.*, *112*(2), 397–404.

Macario, R. (2014). Public–private partnerships in ports: where are we? In: Meersman H, Van de Voorde E, & Vanelslander T (eds) *Port infrastructure finance*. Abingdon: Routledge, pp. 55–68.

Madden, T. J., Scholder Ellen, P. & Ajzen, I. (1992). A comparison of the theory of planned behaviour and the theory of reasoned action. *Pers. Soc. Psychol. Bull.*, *18*, 3–9. http://psp.sagepub.com/content/18/1/3.full.pdf+html.

Mao, Y., & He, J. (2009). Investment decision making for PPP projects based on the black-scholes real option pricing model. In: *2009 International Conference on Management and Service Science*, pp. 1–4.

Marino, G. (1996, October 4), Tangenti per l'alta velocità, coinvolti tutti i partiti, Repubblica.

Marques, R. C., & Berg, S. (2011). Risks, contracts, and private-sector participation in infrastructure. *J. Constr. Eng. Manag.*, *137*(11), 925–932.

Mas-Colell, A., Winston, M., & Green, J. R. (1995). *Microeconomic theory*, New York, NY: Oxford University Press.

Maskin, E. S., & Riley, J. G. (2000). Asymmetric auctions. *Rev. Econ. Stud.*, *67*(232), 413–438.

Masrom, M. A., Skitmore, M., & Bridge, A. (2013). Determinants of contractor satisfaction. *Constr. Manag. Econ.*, *31*(7), 761–779.

Mastrogiacomo, D. (1998, June 25), Tav, indagati eccellenti, Repubblica.

Mauro, P. (1995). Corruption and growth. *Q. J. Econ.*, *110*(3), 681–712. http://doi.org/10.2307/2946696

Mauro, P. (1995). Corruption & growth. *Q. J. Econ.*, *110*(3), 681–712.

Mauro, P. (1998). Corruption and the composition of government expenditure. *J. Public Econ.*, *69*(2), 263–279.

May, D., Wilson, O. D., & Skitmore, M. (2001). Bid cutting: an empirical study of practice in south-East Queensland. *Eng. Constr. Archit. Manag.*, *8*(4), 250–256.

Mazar, N., & Aggarwal, P. (2011). Greasing the palm: Can collectivism promote bribery? *Psychol. Sci.*, *22*(7), 843–848.

Mazar, N., Amir, O., & Ariely, D. (2008b). More ways to cheat expanding the scope of dishonesty. *J. Market. Res.*, *45*(6), 651–653.

McAfee, R. P., & McMillan, J. (1986). Bidding for contracts: a principal-agent analysis. *RAND J. Econ.*, *17*(3), 326.

McAfee, R. P., & McMillan, J. (1988). *Incentive in government contracting.* Toronto, ON: University of Toronto Press.

Meletti, G. (2011, November 22), *Incalza, Il Vero Intoccabile dell'alta velocità* (Il fatto quotidiano).

Merrow, E. W. (2011). Industrial megaprojects: concepts. Hoboken, NJ: Wiley.

Messick, D. M., & Bazerman, M. H. (1996). Ethics for the 21st century: a decision-making approach. *MIT Sloan Manag. Rev.*, *37*(2), 9–22.

Milella, L. (1998, February 8), Alta velocità, arresti a raffica Una cerchia di corruttori, Repubblica.

Milgrom, P. (2004). *Putting auction theory to work.* Cambridge: Cambridge University Press,.

Miller, R., & Lessard, D. R. (2000). *The strategic management of large engineering projects.* Cambridge, MA: MIT Press.

Miltersen, K. R., Sandmann, K., & Sondermann, D. (1997). Closed form solutions for term structure derivatives with log-normal interest rates. *J. Finance*, *52*(1), 409–430.

Min, L. I. U., & Felix, F. W. U. (2009). Wind power investment decision-making strategy based on real options theory. *Autom. Electric Power Syst.*, (21), *33*(21), 19–23.

Ministry of Ecology (2007). Une mobilisation et Une implication des collectivités locales sans précédent. available at: http://archive.wikiwix.com/cache/?url=ht tp://www.equipement.gouv.fr/IMG/pdf/Pages_de_info_fin_cle23fa49.pdf &title=Une%20mobilisation%20et%20une%20implication%20des%20collectivit% EF%BF%BDs%20locales%20sans%20pr%EF%BF%BDc%EF%BF%BDdent.%20Le %20financement%20de%20l?infrastructure,%20sous%20la.

Minnesota Department of Transportation (2016). Online reference. *Official Website of MnDOT.*

Mo, P. H. (2001). Corruption and economic growth. *J. Comp. Econ.*, *29*(1), 66–79.

Mohamed, S., & McCowan, A. K. (2001). Modelling project investment decisions under uncertainty using possibility theory. *Int. J. Proj. Manag.*, *19*(4), 231–241.

Mott MacDonald (2002). Review of large public procurement in the UK. Report to HM Treasury, London, UK.

Mu, R., de Jong, M., & ten Heuvelhof, E. (2010). A typology of strategic behaviour in PPPs for expressways: lessons from China and implications for Europe. *EJTIR*, *10*(1):42–62.

Müller, R., Pemsel, S., & Shao, J. (2015). Organizational enablers for project governance and governmentality in project-based organizations. *Int. J. Proj. Manag.*, *33*(4), 839–851.

Mungiu-Pippidi, A. (2006). Corruption: diagnosis and treatment. *J. Democracy*, *17*, 86–99. http://doi.org/10.1353/jod.2006.0050

Mungiu-Pippidi, A. (2017). The time has come for evidence-based anticorruption. *Nature Human Behav.*, *1*(1), 11. http://doi.org/10.1038/s41562-016-0011

Myerson, R. B. (1977): Graphs and cooperation in games, *Math. Oper. Res.*, *2*, 225–229.

Myerson, R. B. (1991). *Game theory: analysis of conflict.* Cambridge, MA: Harvard University Press, .

NAO (2000). Examining the value for money of deals under the private finance initiative. London, UK: The Stationery Office.

NAO (2003). PFI: construction performance. Report by the Comptroller and Auditor General, HC 371, Session 2002-03, London, UK.

Nash, J. (1950). Equilibrium points in-person games. *Proc. Natl. Acad. Sci.*, *36*(1), 48–49.

Neely, A. (1999). The performance measurement revolution: why now and what next? *Int. J. Oper. Prod. Manag.*, *19*(2), 205–228.

Neely, A., Adams, C., & Crowe, P. (2001). The performance prism in practice. *Meas. Business Excell.*, *5*(2), 6–12.

Neely, A., Adams, C., & Kennerley, M. (2002). *The performance prism: the scorecard for measuring and managing business success.* London, UK: Prentice Hall Financial Times.

Neely, A., Gregory, M., & Platts, K. (2005). Performance measurement system design: a literature review and research agenda. *Int. J. Oper. Prod. Manag.*, *25*(12), 1228–1263.

Ng, W. L. (2008). An efficient and simple model for multiple criteria supplier selection problem. *Eur. J. Oper. Res.*, *186*(3), 1059–1067.

Ng, S. T., Wong, Y. M., & Wong, J. M. (2012). Factors influencing the success of PPP at feasibility stage–A tripartite comparison study in Hong Kong. *Habitat Int.*, *36*(4), 423–432.

Nolan, J. M., Schultz, P. W., Cialdini, R. B., Goldstein, N. J., & Griskevicius, V. (2008). Normative social influence is under-detected. *Personality Social Psychol. Bull.*, *34*(7), 913–923. http://doi.org/10.1177/0146167208316691

Nooteboom, B. (2004). Governance and competence: how can they be combined? *Cambridge J. Econ.*, *28*(4), 505–525.

O'Leary, D., & Stalgren, P (2008). Fighting corruption in water: strategies, tools and ways forward. In: Zinnbauer, D., & Dobson, R (eds). *Global Corruption Report 2008: Corruption in the water sector*, Berlin: Transparency International. Available at: www.transparency.org/publications/gcr

OECD (2006) Mapping out good practices for promoting integrity in public procurement. Available at: www1.fidic.org/resources/integrity/oecd_integ_proc urement_24nov06.doc

Office of Government Commerce (2002). Green public private partnerships. Norwich, UK

Oice (2007). Rotaie verso l'Europa. available at: http://www.oice.it/progettopub blico/31/maire.pdf.

Olivier De Sardan, J. (1999). A moral economy of corruption in Africa? *J. Mod. African Stud.*, 37(1), 25–52.

Olken, B. A., & Pande, R. (2012). Corruption in developing countries. *Annual Rev. Econ.*, 4(1), 479–509. http://doi.org/10.1146/annurev-economics-080511-11 0917

ORGANISATION FOR ECONOMIC COOPERATION AND DEVELOPMENT (1997). Convention on Combating Bribery of Foreign Public Officials in International Business Transactions. Paris: OECD, 1997. See http://www.oecd. org/document/21/0,3343,en_2649_201185_2017813_1_1_1_1,00.html (accessed 25/03/2008).

Ortiz, I. N., Buxbaum, J. N., & Little, R. (2008). Protecting the public interest: role of long-term concession agreements for providing transportation infrastructure. *Transp. Res. Rec.*, 2079(1), 88–95.

Ostrom, E. (2000). Collective action and the evolution of social norms. *J. Econ*, 2079(1), 88–95.

Ostrom, E., Burger, J., & Field, C. (1999). Revisiting the commons: local lessons, global challenges. *Science*, 284(5412), 278–282. http://doi.org/10.1126/sc ience.284.5412.278

Pagano, M. A., & Perry, D. (2008). Financing infrastructure in the 21st century city. *Public Works Manag. Policy*, 13(1), 22–38.

Pakkala, P. (2002). Innovative project delivery methods for infrastructure. Helsinki, Finland: Finnish Road Enterprise.

Panayides P, Parola F, & Lam J. (2015). The effect of institutional factors on public–private partnership success in ports. *Transp. Res. Part A*, 71, 110–127.

Pantelias, A. (2009). A methodological framework for probabilistic evaluation of financial viability of transportation infrastructure under public private partnerships. The University of Texas at Austin.

Pantelias, A., & Zhang, Z. (2010). Methodological framework for evaluation of financial viability of public-private partnerships: investment risk approach. *J. Infrastruct. Syst.*, 16(4), 241–250.

Peirson, G., & McBride, P. (1996). Public/private sector infrastructure arrangements. *CPA Communique*, 73, 1–4.

Percoco, M. (2014). Quality of institutions and private participation in transport infrastructure investment: evidence from developing countries. *Transp. Res. Part A*, 70, 50–58.

Persson, A., Rothstein, B., & Teorell, J. (2012). Why anticorruption reforms fail— Systemic corruption as a collective action problem. *Governance*, 26(3), 449–471. http://doi.org/10.1111/j.1468-0491.2012.01604.x

Pezzella, G. (2011), Lo scandalo della Banca Romana, *Treccani*, available at: http:// www.treccani.it/scuola/lezioni/in_aula/storia/banche/pezzella.html   accessed 13 July 2015.

Phillips, D. (1974). The influence of suggestion on suicide: substantive and theoretical implications of the Werther effect. *Am. Sociol. Rev.*, 39(3), 340–354.

Ping, S. H. (2013). Game theory and PPP. In: *The Routledge Companion to Public-Private Partnerships* (pp. 175–207).

Pinto, J., Leana, C. R., & Pil, F. K. (2008). Corrupt organizations or organizations of corrupt individuals? Two types of organization-level. *Acad. Manag. Rev.*, 33(3), 685–709. http://doi.org./10.5465/AMR.2008.32465726

Pinto, J. K. (2014). Project management, governance, and the normalization of deviance. *Int. J. Proj. Manag.*, *32*(3), 376–387.

Pinto, J. K., & Patanakul, P. (2015). When narcissism drives project champions: a review and research agenda. *Int. J. Proj. Manag.*, *33*(5), 1180–1190.

Pinto, J. K., & Winch, G. (2016). The unsettling of 'settled science:' the past and future of the management of projects. *Int. J. Proj. Manag.*, *34* (2), 237–245.

Piyatrapoomi, N., Kumar, A., & Weligamage, J. (2005). Identification of critical input variables for risk-based cost estimates for road maintenance and rehabilitation. In 4th International Conference on Maintenance and Rehabilitation of Pavements and Technological Control, 2005-08-18–2005-08-20.

Poppo, L., & Zenger, T. (2002). Do formal contracts and relational governance function as substitutes or complements? *Strat. Manag. J.*, *28*(3), 707–725.

Public–Private Infrastructure Advisory Facility (PPIAF). (2015). Online reference. Official website of PPIAF.

Public–Private Infrastructure Advisory Facility. Toolkit for Public– Private Partnership in Roads & Highways. (2016) http://www.ppiaf.org/ppiaf/sites/ppiaf.org/files/documents/toolkits/highwaystoolkit/index.html. Accessed July 28, 2016.

Rai, A. J., Stemmer, P. M., Zhang, Z., Adam, B. L., Morgan, W. T., Caffrey, R. E., ... & Leung, H. C. E. (2005). Analysis of Human Proteome Organization Plasma Proteome Project (HUPO PPP) reference specimens using surface enhanced laser desorption/ionization-time of flight (SELDI-TOF) mass spectrometry: Multi-institution correlation of spectra and identification of biomarkers. *Proteomics*, *5*(13), 3467–3474.

Railway Gazette (2011). *LGV Bretagne Financing Agreement Signed. Railway Gazette* July. (available at: http://www.railwaygazette.com/news/singleview/view/lgv-bretagne-financing-agreement-signed.html.).

Railway Gazette. (2010 March), RFF selects Vinci to build Tours – Bordeaux LGV, Railway Gazette, available at: http://www.railwaygazette.com/news/single-view/view/rff-selects-vinci-to-build-tours-bordeaux-lgv.html.

Raisbeck, P., Duffield, C., & Xu, M. (2010). Comparative performance of PPPs and traditional procurement in Australia. *Constr. Manag. Econ.*, *28*(4), 345–359.

Rakić, B., & Rađenović, T. (2014). Real options methodology in public-private partnership projects valuation. *Econ. Annals*, *59*(200), 91–113.

Rasmusen, E. (2001). *Games and information.* Malden, MA: Blackwell Publisher Inc.

Recarte, P. (2013). Lgv en Service:Modes de Financement Rentabilite, Bilans Ante et Post.

Reinhardt, W. (2011). The role of private investment in meeting US transportation infrastructure needs. *Public Works Financing*, *260*, 1–31.

Reno, R. R., Cialdini, R. B., & Kallgren, C. A. (1993). The trans situational influence of social norms. *J. Person. Social Psychol.*, *64*(1), 104–112. http://doi.org/10.1037//0022-3514.64.1.104

Repubblica (1991. August). *FS, al via i contratti per l'alta velocità*, Repubblica. https://www.repubblica.it/argomenti/Alta_velocit%C3%A0

Repubblica (1994, February). Via all'alta velocità, firmato il contratto per la Roma-napoli, Repubblica. https://www.repubblica.it/viaggi/argomenti/alta_velocit%C3%A0

Repubblica (1998, June). *La Tav napoletana patto per le tangenti*, Repubblica. https://www.repubblica.it/viaggi/argomenti/alta_velocit%C3%A0

Repubblica (2000, August). *Tav, sì alla Milano-Bologna*, Repubblica. https://www.repubblica.it/viaggi/argomenti/alta_velocit%C3%A0

Repubblica (2003, March 19). *Assolti Fusco e Funaro*, Repubblica, available at: https://www.repubblica.it/argomenti/ferrovie_dello_stato.

Repubblica (2013, September 16). *Arrestata Lorenzetti, ex presidente Umbria: accusa di corruzione per Tav in Toscana*, Repubblica, (available at:) https://bari.repubblica.it/cronaca/2021/06/14/news/treni_raddoppio_bari-napoli-306060638/.

RFI (2015). *Rete Alta Velocità - Alta Capacità*. available at: http://www.rfi.it/cms/v/index.jsp?vgnextoid=09dcff14c347b110VgnVCM1000003f16f90aRCRD.

Ridlehoover, J. (2004). Applying Monte Carlo simulation and risk analysis to the facility location problem. *Eng. Econ.*, *49*(3), 237–252.

Robinson, H. S., Anumba, C. J., Carrillo, P. M. & Al-Ghassani, A. M. (2005). Business performance measurement practices in construction engineering organisations. *Meas. Business Excel.*, *9*(1), 13–22.

Roehrich, J. K., Lewis, M. A., & George, G. (2014). Are public–private partnerships a healthy option? A systematic literature review. *Social Sci. Med.*, *113*, 110–119.

Rose-Ackerman, S. (1978). *Corruption: a study in political economy*. New York: Academic Press, pp. 109–135.

Rose-Ackerman, S. (1999). *Corruption and government: causes, consequences and reform*. Cambridge: Cambridge University Press.

Rose-Ackerman, S. (1997). Corruption: causes, consequences, and cures. *Trends in Organized Crime*. http://doi.org/10.1007/s12117-997-1155-3

Rose-Ackerman, S. (1996). Redesigning the state to fight corruption: *transparency, competition and privatization*. Public Policy for the Private Sector, No. 04 (available at: https://openknowledge.worldbank.org/handle/10986/11627\nhttps://openknowledge.worldbank.org/bitstream/10986/11627/1/multi0page.pdf).

Rose-Ackerman, S. (2006). *International handbook on the economics of corruption*. Northampton, MA: Edward Elgar Publishing, Inc.

Rossi, S. (2000, August). *Via libera all'Alta velocità una sfida da 10mila miliardi*, La Repubblica. https://www.repubblica.it/economia/2021/06/15/news/usa_alta_velocita_italo-giapponese_via_al_progetto_del_primo_treno_superveloce-306168540/

Roth, L. (2004). Privatisation of prisons. Background Paper No. 3/04. Sydney, Australia: New South Wales Parliament

Rothchild, M. & Stiglitz, J. (1976). Equilibrium in competitive insurance markets: an essay on the economics of imperfect information. *Q. J. Econ.*, *90*, 629–649.

Rothstein, B. (2000). Trust, social Dilemmas and collective memories. *J. Theor. Politics*, *12*(4), 477–501. http://doi.org/10.1177/0951692800012004007

Rothstein, B. (2011). Corruption: the killing fields quality of government and the health sector. In: Quality of government - corruption, social trust, and inequality in international perspective (pp. 58–76). Chicago, IL: University of Chicago Press.

Rothstein, B., & Eek, D. (2009). Political corruption and social trust: an experimental approach. *Ration. Soc.*, *21*(1), 81–112. http://doi.org/10.1177/1043463108099349

Roumboutsos, A., Farrell, S., Liyanage, C. L., & Macário, R. (2013) COST action TU1001 public private partnerships in transport: trends and theory. *Discussion Papers—Part II* Case Studies. Cost Office 2013, pp. 130–153.

Sachs, T., Elbing, C., Tiong, R. L. K., & Alfen, H. (2005). Efficient assessment of value for money (VFM) for selecting effective public private partnership (PPP) solutions – A comparative study of VFM assessment for PPPs in Singapore and

Germany, Singapore: School for Civil and Environmental Engineering, Nanyang Technological University.

Sachs, T., Tiong, R., & Wang, S. Q. (2007). Analysis of political risks and opportunities in public private partnerships (PPP) in China and selected Asian countries. *Chin. Manag. Stud.*, *1*(2), 126–148. https://doi.org/10.1108/17506140710758026

Sagalyn, L. B. (2007). Public/private development: lessons from history, research, and practice. *J. Am. Plan. Assoc.*, *73*(1), 7–22.

Sargiacomo, M., Ianni, L., D'Andreamatteo, A., & Servalli, S. (2015). Accounting and the fight against corruption in Italian government procurement: a longitudinal critical analysis (1992–2014). *Crit. Perspect. Account.*, *28*, 89–96.

Schmid, A. (2004). *Conflict and cooperation: Institutional and behavioural economics.* Oxford: Blackwell Publishing.

Schultz, P. W., Nolan, J. M., Cialdini, R. B., Goldstein, N. J., & Griskevicius, V. (2007). The constructive, destructive, and reconstructive power of social norms. *Psychol. Sci.*, *18*(5), 429–34. http://doi.org/10.1111/j.1467-9280.2007.01917.x

Schwarz, N., & Strack, F. (1999). Reports of subjective well-being: judgmental processes and their methodological implications. In: D. Kahneman, E. Diener, & N. Schwarz (Eds.), *Well-being: The foundations of hedonic psychology* (pp. 61–84). New York, NY: Russell Sage Foundation Publications.

Scott, W. R. (2005). Institutional theory: contributing to a theoretical research program. In: Smith, K. G., & Hitt, M. A. (Eds.), *Great minds in management: the process of theory development great minds in management: the process of theory development* (pp. 460–484). Oxford: Oxford University Press.

Scott, W. R. (2012). The institutional environment of global project organizations. *Eng. Project Organ. J.*, *2*(1–2), 27–35.

Selvatici, F. (2015). http://ricerca.repubblica.it/repubblica/archivio/repubblica/2015/02/13/lavori-tav-chiesti-39-rinvii-a-giudizioFirenze11.html.

Senato della Repubblica (2007). Indagine conoscitiva sulla situazione economicae finanziaria delle ferrovie dello stato e sullo stato Dei cantieri e Dei costidell'alta velocità ferroviaria. available at: http://www.senato.it/leg/15/BGT/Schede/ProcANL/ProcANLscheda13231.htm.

Seneviratne, P. N., & Ranasinghe, M. (1997). Transportation infrastructure financing: Evaluation of alternatives. *J. Infrastruct. Syst.*, *3*(3), 111–118.

Serra, D., & Wantchekon, L. (2012). Experimental research on corruption: introduction and overview. In: D. Serra & L. Wantchekon (Eds.), *New advances in experimental research on corruption* (pp. 1–11). Emerald Group Publishing Limited. doi:10.1108/S0193-2306(2012)0000015003

Shaoul, J. (2005). A critical financial analysis of the private finance initiative: selecting a financing method or allocating economic wealth? *Critical Perspect. Account.*, *16*, 441–471.

Shen, L. Y., Platten, A., & Deng, X. P. (2006). Role of public private partnerships to manage risks in public sector projects in Hong Kong. *Int. J. Proj. Manag.*, *24*(7), 587–594.

Sheppard, B. M., Hartwick, J. & Warshaw, P. R. (1988). The theory of reasoned action: a meta-analysis of past research with recommendations for modification and future research. *J. Consumer Res.*, *15*, 325–343. http://www.jstor.org/stable/2489467

Shleifer, A., & Vishny, R. W. (1993). Corruption. *Q. J. Econ.*, *108*(3), 599–617.

Siemiatycki, M. (2011). Public-private partnership networks: exploring business-government relationships in United Kingdom transportation projects. *Econ. Geogr.*, *87*(3), 309–334.

Skitmore, M., & Ng, S. T. (2002). Analytical and approximate variance of total project cost. *J. Constr. Eng. Manag.*, *128*(5), 456–460.

Smith, A. J. (1999). *Privatized infrastructure: the role of government*. London: Thomas Telford.

SNFC, L.S.N. des C. de fer F. (2015). *Le financement de la phase 2*. Available at: http:// www.lgv-est.com/decouvrir/le-financement-phase-2/ (accessed 23 September 2015).

Solomon, P. J., & Young, R. R. (2007). *Performance-based earned value*. Hoboken, NY: John Wiley and Sons, Inc.

Song, Y., Wang, X., Wright, G., Thatcher, D., Wu, P., & Felix, P. (2018). Traffic volume prediction with segment-based regression kriging and its implementation in assessing the impact of heavy vehicles. *IEEE Trans. Intell. Transp. Syst.*, *20*(1), 232–243.

Songer, A. D., Diekmann, J., & Pecsok, R. S. (1997). Risk analysis for revenue dependent infrastructure projects. *Constr. Manag. Econ.*, *15*(4), 377–382.

Sonuga, F., Aliboh, O., & Oloke, D. (2002). Particular barriers and issues associated with projects in a developing and emerging economy. Case study of some abandoned water and irrigation projects in Nigeria. *Int. J. Proj. Manag.*, *20*(8), 611–616.

Søreide, T. (2002). Corruption in public procurement Causes, *consequences and cures*. available at: http://www.cmi.no/publications/file/843-corruption-inpublic-procurement-causes.pdf.

Spence, A. M. (1973). Job market signaling. *Q. J. Econ.*, *87*, 355–374.

Spence, A. M. (1974). *Market Signaling: Informational Transfer in Hiring and Related Screening Processes*. Cambridge, MA: Harvard University Press.

Stålgren, P (2006) Corruption in the water sector: causes, consequences and potential reform *Swedish Water House Policy Brief Nr. 4*. Stockholm: SIWI.

Stanghellini, S. (2000). Le système Contractuel Italien en Phase de Changement. Europe : conduite des projets de construction (available at: http://www.chantier. net/europe.html).

Stansbury, N. (2005). Exposing the Foundations of Corruption in Construction. Global Corruption Report *2005*, pp. 36–55.

Stiglitz, J. (2012). *The price of inequality*. London, UK: Penguin.

Tabish, S. Z. S., & Jha, K. N. (2011). Analyses and evaluation of irregularities in public procurement in India. *Constr. Manag. Econ.*, *29*(3), 261–274.

Tabish, S. Z. S., & Jha, K. N. (2012). The impact of anti-corruption strategies on corruption free performance in public construction projects. *Constr. Manag. Econ.*, *30*(1), 21–35.

Tamayo, J., Vassallo, J., & de los Ángeles Baeza, M. (2014). Unbundling tolls from contracts: a new road PPP model. *Public Money Manag.*, *34*(6), 447–451.

Tang, L. Y., Shen, Q., & Cheng, E. (2010). A review of studies on public-private partnership projects in the construction industry. *Int. J. Proj. Manag.*, *28*, 683–694.

Tang, L. C. M., Atkinson, B., & Zou, R. R. (2012). An entropy-based SWOT evaluation process of critical success factors for international market entry: a case study of a medium-sized consulting company. *Constr. Manag. Econ.*, *30*(10), 821–834.

Tanzi, V., & Davoodi, H. (1998). Corruption, public investment, and growth. *The welfare state, public investment, and growth*. Japan: Springer, pp. 41–60.

Tao, S., Wu, C., Sheng, Z., & Wang, X. (2018). Space-time repetitive project scheduling considering location and congestion. *J. Comput. Civil Eng.*, *32*(3), 04018017.

Taylor, K. (2007). Trans Texas Corridor Road Race. *J. New Am.*, *23*(17), 17.

Tervonen, T. (2014). JSMAA: open source software for SMAA computations. *Int. J. Syst. Sci.*, *45*(1), 69–81.

Tervonen, T., & Lahdelma, R. (2007). Implementing stochastic multicriteria acceptability analysis. *Eur. J. Oper. Res.*, *178*(2), 500–513.

Tervonen, T., Hakonen, H., & Lahdelma, R. (2008). Elevator planning with stochastic multicriteria acceptability analysis. *Omega*, *36*(3), 352–362.

Thaler, R., & Sunstein, C. (2008). *Nudge: Improving decisions about health, wealth, and happiness*. New Haven, CT: Yale University Press.

Thobani, M. (1998). *Private Infrastructure, Public Risk*.

Thomas Ng S., Wong Y., & Wong J. (2012) Factors influencing the success of PPP at feasibility stage: a tripartite comparison study in Hong Kong. *Habitat Int.*, *36*, 423–432.

Tiong, R. L. K. (1995). Competitive advantage of equity in BOT tender. *J. Constr. Eng. Manag.*, ASCE, *121*, 282–288.

Tiong, R. L. K. (1996). CSFs in competitive tendering and negotiation model for BOT projects. *J. Constr. Eng. Manag.*, ASCE, *122*(3), 205–211.

Torra, V. (2010). Hesitant fuzzy sets. *Int. J. Intell. Syst.*, *25*(6), 529–539.

Transparency International (2005). Global Corruption Report 2005: Corruption in construction and postconflict reconstruction. Available at: www.transparency.org/publications/gcr

Transparency International (2011). Bribe payer index 2011. available at: http://www.transparency.org/bpi2011.

Transparency International (2013). National Integrity System Assessment. available at: https://www.transparency.org/whatwedo/nis.

Transparency International (2014). Corruption perceptions indeex 2014: results. available at: https://www.transparency.org/cpi2014/results.

Transparency International (2015a). FAQS on corruption. available at: https://www.transparency.org/whoweare/organisation/faqs_on_corruption/2/ (accessed 9 July 2015).

Transparency International (2015b). National integrity system assessments. available at: https://www.transparency.org/whatwedo/nis.

Transparency International (2002). Bribe Payers Index 2002. TI, Berlin. http://www.transparency.org/policy_research/surveys_indices/bpi.

Transparency International (2005). Global Corruption Report-2005: Corruption in Construction and Post-Conflict Reconstruction, at http://www.transparency.org/publications/gcr.

Transparency International (2005). Global Corruption Report-2005. London: Pluto Press, 2005.

Treasury Taskforce (1998). Partnerships for prosperity – The Private Finance Initiative. London, UK: HM Treasury.

Treasury Taskforce (1999). Step-by-step guide to the PFI procurement process. <http://www.treasury-projects.gov.uk> (Accessed 8 Aug. 2012).

Treisman, D. (2000). The causes of corruption: a cross-national study. *J. Public Econ.*, *76.* http://doi.org/10.1016/S0047-2727(99)00092-4

Treisman, D. (2007). What have We learned about the causes of corruption from ten years of cross-National Empirical Research? *Annu. Rev. Poli. Sci.*, *10* (1), 211–244.

Trenitalia (2015). Biglietti. available at: http://www.trenitalia.com/.

Trigeorgis, L. (1996). *Real options: Managerial flexibility and strategy in resource allocation.* Cambridge, MA: MIT Press.

Tsamboulas, D., Verma, A., & Moraiti, P. (2013) Transport infrastructure provision and operations: why should governments choose private-public partnership? *Res. Transp. Econ.*, *38*:122–127

Turner, R., & Zolin, R. (2012). Forecasting success on large projects: developing reliable scales to predict multiple perspectives by multiple stakeholders over multiple time frames. *Proj. Manag. J.*, *43* (5), 87–99.

Turner, S. M., Eisele, W. L., Benz, R. J., & Holdener, D. J. (1998). *Travel time data collection handbook (No. FHWA-PL-98-035).* United States. Federal Highway Administration.

U.S. Department of Transportation (2007). Case studies of transportation public-private partnerships around the world. Final Report. Prepared for Office of Policy and Governmental Affairs. July, 2007. 3–7.

Uberti, L. J. (2016). Can institutional reforms reduce corruption? Economic theory and patron–client politics in developing countries. *Dev. Chang.*, *47*(2), 317–345.

Vajdić, N., & Damnjanović, I. (2011, June). Risk management in public-private partnership road projects using the real options theory. In: International Symposium Engineering Management and Competitiveness (EMC2011), June 24–25, 2011, Zrenjanin, Serbia.

Van de Graaf, T., & Sovacool, B. K. (2014). Thinking big: politics, progress, and security in the management of Asian and European energy megaprojects. *Energy Policy*, *74*, 16–27.

van Haastrecht, A., & Pelsser, A. (2011). Generic pricing of FX, inflation and stock options under stochastic interest rates and stochastic volatility. *Quant. Finance*, *11*(5), 665–691.

Van Veldhuizen, R. (2013). The influence of wages on public officials' corruptibility: a laboratory investigation. *J. Econ. Psychol.*, *39*, 341–356. http://doi.org/10.10 16/j.joep.2013.09.009

Vandoros, N., & Pantouvakis, J. P. (2006, November). Using real options in evaluating PPP/PFI projects. In Symposium on sustainability and value through construction procurement (p. 594).

Vannucci, A. (2009). The controversial legacy of 'Mani pulite': a critical analysis of Italian corruption and anti-corruption policies. Bull. *Italian Polit.*, *1*(2), 233–264.

Vassallo, J. M. (2006). Traffic risk mitigation in highway concession projects: the experience of Chile. *J. Transp. Econ. Policy*, *40*(3), 359–381.

Vee, C., Skitmore, C. (2003). professional ethics in the construction industry, *Eng.. Constr. Architect. Manag.*, *10*(2), 117–127.

Verhoest K, Petersen O, Scherrer W, Soecipto R (2015) How do governments support the development of public private partnerships? Measuring and comparing PPP governmental support in 20 European countries. *Transp. Rev. Transnatl. Transdiscipl. J.*, *35*(2), 118–139.

Victorian Audit-General's Office (2002). Report on public sector agencies. Melbourne, Australia: Victorian Auditor-General's Office.

Vincent-Jones, P. (1997). Hybrid organisation, contractual governance, and compulsory competitive tendering in the provision of local authority services. In: Deakin, S., & Michie, J. (Eds.), Contracts, co-operation and competition (pp. 143–174). Oxford: Oxford University Press.

Vinci, E. (2003, January 31). *Tangenti, da rifare il processo Necci*, Repubblica. https://www.repubblica.it/online/fatti/necci/arresto/arresto.html?ref=search

Von Neumann, J. (1966). In: Theory of self-reproducing automata (A. W. Burk, ed.). Urbana, IL: University of Illinois Press. https://cba.mit.edu/events/03.11.ASE/docs/VonNeumann.pdf

Wacker, R. R., & Roberto, K. A. (2010). *Aging social policies: an international perspective*. London, UK: Sage Publications.

Walker, C., & Smith, A. J. (1995). *Privatized infrastructure- the BOT approach*. New York, NY: Thomas Telford Inc.

Wall, D. M. (1997). Distributions and correlations in Monte Carlo simulation. *Constr. Manag. Econ.*, *15*(3), 241–258.

Walsh, K. (1995). *Public Services and Market Mechanisms*. London: Macmillan.

Wan, S. P., & Li, D. F. (2013). Atanassov's intuitionistic fuzzy programming method for heterogeneous multiattribute group decision making with Atanassov's intuitionistic fuzzy truth degrees. *IEEE Trans. Fuzzy Syst.*, *22*(2), 300–312.

Wang, D., Zhang, L., Ma, N., & Li, X. (2007). Two secret sharing schemes based on Boolean operations. *Pattern Recognit.*, *40*(10), 2776–2785.

Wang, P., Shen, J., & Zhang, B. (2016). A new method for two-sided matching decision making of PPP projects based on intuitionistic fuzzy choquet integral. *J. Intell. Fuzzy Syst.*, *31*(4), 2221–2230.

Wang, S. Q., Tiong, R. L., Ting, S. K., & Ashley, D. (2000). Evaluation and management of political risks in China's BOT projects. *J. Constr. Eng. Manag.*, *126*(3), 242–250.

Wedeman, A. (1997). Lootcers, and dividend-ollectors: corruption and growth in Zaire, South Korea, and the Philippines. *J. Dev. Areas*, *31*(4), 457–478.

Wei, S.-J. (2000). How taxing is corruption on international investors? *Rev. Econ. Stat.*, *82*(1), 1–11.

Weisel, O., & Shalvi, S. (2015). The collaborative roots of corruption. *Proc. Natl. Acad. Sci.*, *112*(34), 10651–10656. http://doi.org/10.1073/pnas.1423035112

Wells, J. (2014). Corruption and collusion in construction: a view from the industry. In: Søreide, T., Aled, W. (Eds.), *Corruption, grabbing and development. real world challenges* (pp. 23–34). Cheltenham, UK: Edward Elgar Publishing Ltd.

Welsh, D. T., Ordóñez, L. D., Snyder, D. G., & Christian, M. S. (2014). The slippery slope: how small ethical transgressions pave the way for larger future transgressions. *J. Appl. Psychol.*, *99*, 114–127. doi:10.1037/a0036950

Wenzel, M. (2004). An analysis of norm processes in tax compliance. *J. Econ. Psychol.*, *25*(2), 213–228. http://doi.org/10.1016/S0167-4870(02)00168-X

Wibowo, A., & Alfen, H. W. (2015). Government-led critical success factors in PPP infrastructure development. *Built Environ. Project Asset Manage.*, *5*(1), 121–134. https://doi.org/10.1108/BEPAM-03-2014-0016

Williams, C. C., Horodnic, I. A., & Windebank, J. (2015). Explaining participation in the informal economy: an institutional incongruence perspective. *Int. Sociol.*, *30*(3), 294–313.

Williamson, O. E. (1996). *The Mechanics of Governance.* New York: Oxford University Press.

Willoughby, C. (2013) How much can public private partnership really do for urban transport in developing countries? *Res. Transp. Econ., 40,* 34–55.

Winch, G. (2000a). *Editorial construction business systems in the European Union. Build. Res. Inf. 28*(2), 88–97.

Winch, G. (2000b). Institutional reform in British construction: partnering and private finance. *Build. Res. Inf. 28*(2), 141–155.

Winch, G., & Leiringer, R. (2016). Owner project capabilities for infrastructure development: a review and development of the 'strong owner' concept. *Int. J. Proj. Manag., 34*(2), 271–281.

Winch, G. M. (2013). Escalation in major projects: lessons from the channel fixed link. *Int. J. Proj. Manag., 31*(5), 724–734.

Winch, G. M. (2014). Three domains of project organising. *Int. J. Proj. Manag., 32*(5), 721–731.

World Bank (2004). *Water resources sector strategy: strategic directions for World Bank engagement.* Washington, DC: World Bank.

World Bank (2013). Worldwide governance indicator. available at: http://info.worldbank.org/governance/wgi/index.aspx.

World Commission on Dams (2000). *Dams and development: a new framework for decision-making,* London: Earthscan.

World Economic Forum (2015). Global competitiveness ranking: ethics and corruption. available at: http://reports.weforum.org/global-competitivenessreport-2014-2015/rankings/.

World Water Assessment Programme (2009). The United Nations World Water Development Report 3: Water in a Changing World, Paris: UNESCO and London: Earthscan.

Wright, T. (1997). *Variability in traffic monitoring data.* https://doi.org/10.2172/629487

Xenidis, Y., & Angelides, D. (2005). The financial risks in build-operate-transfer projects. *Constr. Manag. Econ., 23*(4), 431–441.

Xia, M., & Xu, Z. (2011). Hesitant fuzzy information aggregation in decision making. *Int. J. Approximate Reason., 52*(3), 395–407.

Xia, W., & Wu, Z. (2007). Supplier selection with multiple criteria in volume discount environments. *Omega, 35*(5), 494–504.

Xiaohu, Z., Heydari, M., Lai, K. K., & Yuxi, Z. (2020). Analysis and modeling of corruption among entrepreneurs. *REICE: Revista Electrónica de Investigación en Ciencias Económicas, 8*(16), 262–311.

Xie, J., & Thomas Ng, S. (2013). Multiobjective Bayesian network model for public-private partnership decision support. *J. Constr. Eng. Manag., 139*(9), 1069–1081.

Xu, Y. (2010). Developing a risk assessment model for PPP projects in China—a fuzzy synthetic evaluation approach. *Autom. Constr., 19,* 929–943.

Yang, F., Ang, S., Xia, Q., & Yang, C. (2012). Ranking DMUs by using interval DEA cross efficiency matrix with acceptability analysis. *Eur. J. Oper. Res., 223*(2), 483–488.

Yang, F., Song, S., Huang, W., & Xia, Q. (2015). SMAA-PO: project portfolio optimization problems based on stochastic multicriteria acceptability analysis. *Annals Oper. Res., 233*(1), 535–547.

Ye, S., & Tiong, R. L. (2000). NPV-at-risk method in infrastructure project investment evaluation. *J. Constr. Eng. Manag., 126*(3), 227–233.

Yescombe, E. R. (2011). *Public-private partnerships: principles of policy and finance.* London, UK: Elsevier.

Yin, R. K. (2013). *Case study research: design and methods.* London, UK: SAGE Publications.

Yolles, M. (2009). A social psychological basis of corruption and sociopathology. *J. Organ. Change Manag.*, Volume 22, pp. 691–731. http://www.emeraldinsight.com/journals.htm?issn=0953 4814&volume=22&issue=6&articleid=1817029&show=html

Yong, H. K. (2010). *Public-private partnerships policy and practice.* London, UK: Commonwealth Secretariat.

Yuan, J., Zeng, A. Y., Skibniewski, M. J., & Li, Q. (2009). Selection of performance objectives and key performance indicators in public-private partnership projects to achieve value for money. *Constr. Manag. Econ.*, *27*(3), 253–270.

Yun, S., Jung, W., Han, S. H., & Park, H. (2015) Critical organizational success factors for public private partnership projects—a comparison of solicited and unsolicited proposals. *J. Civil Eng. Manag.*, *21*(2), 131–143.

Zarkada-Fraser, A., & Skitmore, M. (2000). Decisions with moral content: collusion. *Constr. Manag. Econ.*, *18*(1), 101–111.

Zhang, Q., Lai, K. K., & Yen, J. (2019). Multicriteria supplier selection using acceptability analysis. *Adv. Mech. Eng.*, *11*(10), 1687814019883716.

Zhang, S., Chan, A. P., Feng, Y., Duan, H., & Ke, Y. (2016). Critical review on PPP Research–A search from the Chinese and International Journals. *Int. J. Proj. Manag.*, *34*(4), 597–612.

Zhang, X., Bao, H., & Skitmore, M. (2015). The land hoarding and land inspector dilemma in China: an evolutionary game theoretic perspective. *Habitat Int.*, *46*, 187–195.

Zheng, J., Roehrich, J. K., & Lewis, M. A. (2008). The dynamics of contractual and relational governance: evidence from long-term public-private procurement arrangements. *J. Purchas. Supply Manag.*, *14*(1), 43–54.

Zheng, S., Xu, K., He, Q., Fang, S., & Zhang, L. (2018). Investigating the sustainability performance of ppp-type infrastructure projects: a case of China. *Sustainability*, *10*(11), 4162.

Zietlow, G. (2005). Cutting costs and improving quality through performance-based road management and maintenance contracts – The Latin American and OECD experience. Birmingham, 24–29 April, University of Birmingham (UK), Senior Road Executives Programme, Restructuring Road Management.

Zou, W., Kumaraswamy, M., Chung, J., & Wong, J. (2014). Identifying the critical success factors for relationship management in PPP projects. *Int. J. Proj. Manag.*, *32*(2), 265–274.

Index

Printed in the United States
by Baker & Taylor Publisher Services